Wolfgang Merbach, Birgit W. Hütsch,
Lutz Wittenmayer, Jürgen Augustin (Hrsg.)

Prozessregulation in der Rhizosphäre

Borkheider Seminare zur Ökophysiologie des Wurzelraumes

Der Pflanzenbewuchs, das dazugehörige Wurzelsystem und der durchwurzelte Bodenraum nehmen eine Schlüsselstellung in terrestrischen Ökosystemen ein. Hier vollziehen sich komplizierte Wechselwirkungen zwischen Pflanzenstoffwechsel und Umweltfaktoren einerseits und (angetrieben durch die C-Lieferung der Pflanzen) zwischen Pflanzenwurzeln, Mikroben, Bodentieren, organischen C- und N-Verbindungen sowie mineralischen Bodenbestandteilen andererseits. Diese haben entscheidende Bedeutung für die Pflanzen- und Bodenentwicklung, die Nettostoff- und Nettoenergieflüsse sowie für die Belastungstoleranz von Pflanzen und Ökosystemen. Ihr Verständnis ist daher eine Voraussetzung für die Prognose, Abpufferung und Indikation von Umweltbelastungen, die Berechnung von Stoffflüssen sowie für ökologisch ausgerichtete Regulationsinstrumentarien. Trotz vieler Einzelkenntnisse sind aber derzeit Wirkungsgefüge und Regulationsmechanismen im Pflanze-Boden-Kontaktraum nur ungenügend bekannt, da in den meisten bisherigen Forschungsansätzen dieser Bereich als „Nebeneinander" von Einzelelementen (z. B. von Strukturelementen, Nettostoffflüssen zwischen Grenzflächen, Biozönosepartnern) betrachtet wurde und kaum als Netzwerk funktionaler Kompartimente wechselnder Zusammensetzung. Abhilfe kann hier nur eine systemare Betrachtungsweise der Pflanze-Boden-Wechselbeziehungen auf der Basis einer langfristig und interdisziplinär angelegten ökophysiologischen Forschung schaffen, die auf die Aufklärung der mikrobiologischen, physiologischen, (bio)chemischen und genetischen Interaktionen im System Pflanze – Wurzel – Boden in Abhängigkeit von natürlichen und anthropogenen Einflußfaktoren ausgerichtet ist.

Die 1990 von der Deutschen Landakademie Borkheide (Krs. Potsdam-Mittelmark) und dem heutigen Institut für Primärproduktion und Mikrobielle Ökologie des Zentrums für Agrarlandschafts- und Landnutzungsforschung (ZALF) Müncheberg ins Leben gerufenen Borkheider Seminare zur Ökophysiologie des Wurzelraumes wollen daher Wissenschaftler unterschiedlicher Fachgebiete mit dem Ziel zusammenführen, experimentelle Ergebnisse ohne Zeitdruck zu diskutieren und die Forschung enger zu verflechten. Das unveränderte Interesse an der Tagungsreihe – sie hat 2002 bereits das 13. Mal stattgefunden – spricht für sich selbst. Nachdem die ersten vier Tagungsbände (1990 bis 1993) im Selbstverlag herausgegeben wurden, hat seit dem 5. Band (Mikroökologische Prozesse im System Pflanze – Boden) der Teubner-Verlag diese Aufgabe übernommen. Dafür gebührt ihm der Dank der Herausgeber.

Wolfgang Merbach

Wolfgang Merbach, Birgit W. Hütsch,
Lutz Wittenmayer, Jürgen Augustin (Hrsg.)

Prozessregulation in der Rhizosphäre

13. Borkheider Seminar zur Ökophysiologie des Wurzelraumes

Wissenschaftliche Arbeitstagung in Schmerwitz/Brandenburg
vom 19. bis 20. September 2002

Teubner

Springer Fachmedien Wiesbaden GmbH

Bibliografische Information der Deutschen Bibliothek
Die Deutsche Bibliothek verzeichnet diese Publikation in der Deutschen Nationalbibliographie; detaillierte bibliografische Daten sind im Internet über <http://dnb.ddb.de> abrufbar.

Die Beiträge dieses Bandes wurden von Mitgliedern der Deutschen Gesellschaft für Pflanzenernährung sowie der Kommission IV der Deutschen Bodenkundlichen Gesellschaft begutachtet.

Prof. Dr. habil. Wolfgang Merbach
Geboren 1939 in Ranis (Thüringen). 1958 bis 1964 Landwirtschaftsstudium, 1965 bis 1966 Chemiestudium, 1970 Promotion Universität Jena. 1982 Habilitation Martin-Luther-Universität Halle-Wittenberg (MLU). 1986 bis 1990 Leiter Isotopenlabor Forschungszentrum Bodenfruchtbarkeit Müncheberg, 1989/90 Leiter der AG „Ökologischer Umbau" und Mitglied des zentralen „Runden Tisches" der DDR in Berlin, 1990 Professor Akademie der Landwirtschaftswiss., 1992 bis 1998 Institutsleiter und stellv. Direktor Zentrum für Agrarlandschafts- und Landnutzungsforschung (ZALF) Müncheberg. Seit 1998 Professor für Physiologie und Ernährung der Pflanzen, seit 2000 Dekan Landwirtschaftliche Fakultät der MLU. Vorlesungen Pflanzenernährung, Düngung, Ökotoxikologie, Bodenkunde Universitäten Halle, Jena, Potsdam, Cottbus. Arbeitsschwerpunkte: Symbiontische N_2-Fixierung, Ökophysiologie, Stoffumsatz in der Rhizosphäre, Lachgasemission aus Niedermooren, N-Umsatz in Ökosystemen. Über 250 Publikationen, Herausgeber zahlreicher Bücher. Mitglied in mehreren Editorial Boards und Fachgesellschaften, 1.Vorsitzender Deutsche Gesellschaft für Pflanzenernährung (1997-2001), Mitglied im Council International Ecological Centre, Polnische Akademie der Wissenschaften.

Priv.-Doz. Dr. habil. Birgit W. Hütsch
Geboren 1961 in Fulda (Hessen). 1981 bis 1986 Studium der Agrarwissenschaften, 1991 Promotion Justus-Liebig-Universität (JLU) Gießen, 1992 bis 1993 Rothamsted Experimental Station (UK), 1993 bis 2001 wiss. Mitarbeiterin, 1999 Habilitation JLU Gießen, seit 2001 wiss. Mitarbeiterin, Vertretungsprofessorin für Bodenbiologie und Bodenökologie MLU. Vorlesungen Pflanzenernährung, Bodenökologie Universitäten Gießen, Halle. Arbeitsschwerpunkte: Emission klimarelevanter Spurengase, Rhizodeposition.

Dr. Lutz Wittenmayer
Geboren 1961 in Sondershausen (Thüringen). 1980 bis 1985 Studium der Landwirtschaft und Pflanzenzüchtung, Timirjasew-Akademie Moskau, 1991 Promotion MLU, seit 1996 wiss. Mitarbeiter Institut für Bodenkunde u. Pflanzenernährung MLU. Arbeitsschwerpunkte: Pflanzenstreß, Phytohormone und Wurzelexsudation.

Dr. Jürgen Augustin
Geboren 1954 in Ostritz (Sachsen). 1975 bis 1979 Studium der Pflanzenproduktion und Biochemie, 1985 Promotion MLU, seit 1985 wiss. Mitarbeiter Forschungszentrum Bodenfruchtbarkeit, 1992 bis 1998 am ZALF. 1998 bis 1999 komm. Institutsleiter Inst. für Rhizosphärenforschung u. Pflanzenernährung, danach Abteilungsleiter Inst. für Primärproduktion u. Mikrobielle Ökologie ZALF. Lehraufträge Ökotoxikologie FH Eberswalde, BTU Cottbus. Arbeitsschwerpunkte: Stoffumsatz und Spurengasemission in Feuchtgebieten.

1. Auflage August 2003
ISBN 978-3-519-00447-9 ISBN 978-3-663-07809-8 (eBook)
DOI 10.1007/978-3-663-07809-8

Ursprünglich erschienen bei B.G. Teubner Verlag / GWV Fachverlage GmbH, Wiesbaden 2003.

Umschlaggestaltung: Ulrike Weigel, www.CorporateDesignGroup.de

Gedruckt auf säurefreiem und chlorfrei gebleichtem Papier.

Vorwort

Der vorliegende Band enthält die Kurzfassungen der Vorträge, die zum 13. Borkheider Seminar zur Ökophysiologie des Wurzelraumes, das am 19. und 20. September 2002 in Schmerwitz (Kreis Potsdam-Mittelmark, Brandenburg) stattfand, gehalten wurden. Dabei handelt es sich um Beiträge zu folgenden Themenkreisen:

1. Das Wurzelwachstum und seine Beeinflussung mit den Schwerpunkten Ektomykorrhiza, Feinwurzelbildung und Stickstoffeffizienz (fünf Beiträge)

2. Mikroökologische und molekularbiologische Wechselwirkungen in der Rhizosphäre (sechs Beiträge)

3. Stofftransformation und -akkumulation im Wurzelraum (drei Beiträge)

4. Wirkungen wurzelbürtiger Verbindungen (zwei Beiträge) und

5. Kohlenstoff- und Stickstoffumsatz im durchwurzelten Bodenraum (vier Beiträge).

Auch an diesem Workshop nahmen viele junge Wissenschaftler teil und trugen experimentelle Ergebnisse aus der Sicht der unterschiedlichsten Fachdisziplinen bei. So wurden u.a. bodenkundliche, mikrobiologische, physiologische, biochemische, pflanzenbauliche und pflanzenernährerische Arbeiten dargeboten. Dies ermöglichte eine interdisziplinäre Diskussion zur Vernetzung und Regulation der Einzelprozesse im Sinne des besseren Verständnisses des mikroökosystemaren Wirkungsgefüges im Pflanze-Wurzel-Boden-Kontaktraum.
Eine derartige Betrachtungsweise lässt Impulse für die Entwicklung standort- und umweltgerechter Landnutzungssysteme, die ökophysiologische Indikation von Standorten und Bewirtschaftungsformen, die Aufklärung und Quantifizierung von pflanzenbewuchsvermittelten Stoffbilanzgrößen sowie die Voraussage der Stressbelastbarkeit von Pflanzen und Pflanzengesellschaften erwarten. Sie könnte somit dazu beitragen, die in Ökosystemen ablaufenden Vorgänge besser zu verstehen und die nachhaltige umweltgerechte Gestaltung unserer Kulturlandschaft nutzbar zu machen.

Wissenschaftlicher und organisatorischer Träger des Seminars war wie in den Vorjahren die Professur „Physiologie und Ernährung der Pflanzen“ der Landwirtschaftlichen Fakultät der Martin-Luther-Universität Halle-Wittenberg. Weiterhin waren das Institut für Primärproduktion und Mikrobielle Ökologie im Zentrum für Agrarlandschafts- und Landnutzungsforschung (ZALF) Müncheberg (Kreis Märkisch-Oderland, Brandenburg), die Deutsche Gesellschaft für Pflanzenernährung und die Kommission IV (Bodenfruchtbarkeit und Pflanzenernährung) der Deutschen Bodenkundlichen Gesellschaft (DBG) an der Ausrichtung des Seminars beteiligt. Gastgeber war das Seminar- und Tagungszentrum Schmerwitz, das die Tagungsräume, die Vorführtechnik, die Unterbringung und Verpflegung sicherstellte.
Wir sind der Eigentümerin des Zentrums, Frau Morgenstern, und dem Geschäftsführer, Herrn Rost, zu Dank verpflichtet. Ferner danken wir Herrn Dr. Wolfgang Gans, Institut für Bodenkunde und Pflanzenernährung der Universität Halle-Wittenberg, für die organisatorische Vorbereitung, Frau Dr. Heidrun Beschow für die redaktionelle Bearbeitung des Tagungsbandes und Herrn Jürgen Weiß vom Teubner-Verlag für die gute Zusammenarbeit.

Halle und Müncheberg,
im Juni 2003

Wolfgang Merbach
Birgit W. Hütsch
Lutz Wittenmayer
Jürgen Augustin

Inhaltsverzeichnis

1 Einflussfaktoren des Wurzelwachstums

2 Mikroökologische und molekularbiologische Wechselwirkungen in der Rhizosphäre

3 Stofftransformation und -akkumulation im Wurzelraum

4 Wirkungen wurzelbürtiger Verbindungen

5 Kohlenstoff- und Stickstoffumsatz im durchwurzelten Bodenraum

1

Einflussfaktoren des Wurzelwachstums

Prozessregulation in der Rhizosphäre
13. Borkheider Seminar zur Ökophysiologie des Wurzelraumes
Hrsg.: W. Merbach, B. W. Hütsch, L. Wittenmayer, J. Augustin
B. G. Teubner Stuttgart, Leipzig, Wiesbaden (2003), S. 13-19

Morphologische Interaktionen und molekulare Regulation vor und während der Ektomykorrhiza-Bildung zwischen *Quercus robur* L. und *Piloderma croceum*

Sylvie HERRMANN[1,2], Andrea KRÜGER[1,2], Tatjana PEŠKAN[2], Ralf OELMÜLLER[2], François BUSCOT[1]
[1] Institut für Ökologie, Lehrbereich Umweltwissenschaften
[2] Institut für Botanik, Lehrstuhl Pflanzenphysiologie
Friedrich-Schiller-Universität Jena, Dornburger Str. 159, D-07743 Jena

Abstract

In order to study the events occurring before mycorrhization between oaks (*Quercus robur* L.) and the ectomycorrhizal fungus *Piloderma croceum*, oak seedling morphogenesis was highlighted as basis to characterize the developmental behaviour of micro-propagated oaks in a semi-sterile Petri-dish mycorrhization system. As well as in seedlings, shoots of the micro-cuttings remained characterized be an endogen rhythmic growth under controlled culture conditions (25°C and 16 h illumination). Lateral roots of the micro-propagated oaks also displayed a rhythmic growth, whereby the root and shoot flushes were out of phase, with concomitance of apical bud swelling and maximal growth rate of the roots. In micro-cuttings inoculated with *P. croceum*, this concomitance was maintained, in spite of a strongly enhanced root and shoot growth induced by the fungus in its long pre-mycorrhizal stage. The specificity of morphological modifications (for example enhanced leaf area and lateral root growth) during the prolonged pre-mycorrhizal stage was demonstrated using activated charcoal and auxine (IAA). Both compounds simultaneously advanced mycorrhizal formation and stopped the enhanced development of first order mother roots. In order to elucidate molecular mechanisms involved during this prolonged pre-mycorrhizal stage, a subtractive cDNA bank was established between roots sampled before mycorrhization and control roots mixed with fungal mycelium. After southern blot analysis 60 differentially expressed genes were sequenced and homology was determined for about 30 of them. Northern blot analysis revealed an up regulation during the pre-mycorrhizal stage of two genes encoding for metallothionein and formate-dydrogenase, respectively. The model will serve for further analyses of gen regulation at different stages of mycorrhiza formation.

Einleitung

Ektomykorrhiza (EM)-Pilze haben einen starken Einfluss auf die Wurzelmorphologie. Die dichotome Verzweigung von Kiefernwurzeln unter dem Einfluss von EM-Pilzexsudaten ist ein klassisches Beispiel dafür (SLANKIS 1948), ebenso die allgemeine Verkürzung von Feinwurzeln im Zusammenhang mit der EM-Bildung. Weitere morphogene Einflüsse von EM-Pilzen, insbesondere vor der Mykorrhizierung, wurden dahingegen seltener beschrieben und berücksichtigt.
In einem Modellsystem zur Dualkultur von Stieleichen (*Quercus robur* L.) und dem Basidiomyceten *Piloderma croceum* konnten deutliche morphologische Veränderungen sowohl im Spross als auch im Wurzelbereich vor der Mykorrhizierung festgestellt werden (HERRMANN *et al.* 1998). Diese Effekte wurden im Zusammenhang mit Vorkenntnissen im Bereich der Architektur und Morphogenese der Stieleiche analysiert, um das System genauer zu charakterisieren und es für Untersuchungen der Frühphase der Bildung von EM-Symbiosen auf molekularer Ebene zu nutzen.

Ergebnisse

A. Architektur und Morphologie der Eiche

Unter gemäßigten Klimaten und insbesondere in Europa entwickeln sich Bäume überwiegend nach zwei definierten Architekturmodellen: zum einen nach dem von *Rauh*, zum anderen nach dem von *Troll*. Beide Modelle sind durch ein unendliches, rhythmisches Wachstum der morphologischen Einheiten (Triebe) charakterisiert (HALLÉ und OLDEMAN 1970), das den Bäumen ihr hohes Verzweigungspotential und somit bessere Überlebenschancen unter gemäßigten Klimaten gewährt. In Kulturen von Eichensämlingen unter optimalen und konstanten Wachstumsbedingungen (25°C und 16 h Licht) konnte gezeigt werden, dass dieses unendliche, rhythmische Wachstum endogen bestimmt ist (KLEBS 1911). Die Entwicklung von Wachstumseinheiten mit Bildung von Blättern und Internodien unterschiedlicher Längen bzw. Flächen stellen den räumlichen Ausdruck des rhythmischen Wachstums dar (ALATOU *et al.* 1989). Unter natürlichen klimatischen Bedingungen mit entsprechenden jahreszeitlichen Variationen bilden sich lediglich im Frühling, ggf. auch im Herbst 2 bis 3 Wachstumseinheiten pro Jahr. Unter konstant optimalen Kulturbedingungen werden hingegen alle 3 Wochen neue Sprosswachstumsschübe induziert.
Das Wurzelsystem von Eichensämlingen ist durch eine sehr ausgeprägte Hierarchie charakterisiert (RIEDACKER und BELGRAND 1983). Es besteht aus einer Pfahlwurzel mit senkrechtem (orthogeotropem), schnellem und unbegrenztem Wachstum und Seitenwurzeln mit waagerechtem (plagiotropem), langsamem und begrenztem Wachstum. Gleichzeitig zeigt dieses Wurzelsystem eine hohe Plasti-

zität in seiner Morphologie. Nach Umgehen von Hindernissen findet z.B. die Pfahlwurzel ihr senkrechtes Wachstum wieder (RIEDACKER *et al.* 1981). Die starke Seitenwurzelentwicklung bei Stecklingen aus alten Stumpftrieben und an Hauptwurzeln, die in hydromorphen Böden Wasserlinsen antreffen, stellen weitere Beispiele dieser Plastizität dar (RIEDACKER und BELGRAND 1983).
Untersuchungen der Wurzel-Morphogenese an Eichen ergeben kein einheitliches Bild bezüglich der Wachstumsdynamik. Während bei Pfahlwurzeln ein lineares Wachstum allgemein beschrieben wird, wurden bei Seitenwurzeln sowohl ein lineares (an jungen Sämlingen) als auch ein rhythmisches (an 2 Jahre alten Sämlingen) Wachstum festgestellt (PARMENTIER *et al.* 1991, REICH *et al.* 1980). Das rhythmische Wachstum der Seitenwurzeln scheint also erst bei älteren Sämlingen einzutreten.

B. Verhalten von Eichenstecklingen in einem semi-sterilen *In-vitro*-Kultursystem

Vor dem Hintergrund der o.g. Komplexität der Eichenmorphogenese musste bei der Verwendung von *in vitro* vermehrten Eichenstecklingen zur Untersuchung des morphologischen Einflusses von EM-Pilzen das Verhalten dieser Stecklinge vorab charakterisiert werden.
Der Eichenklon, der für unsere Versuche verwendet wird, wurde vor ca. 15 Jahren aus einem Sämling gewonnen. Die Vermehrung der Sprosse findet in Abständen von 6 Wochen statt. Im Anschluss an diese Vermehrungsphase wird die Bewurzelung auf einem Medium mit Aktivkohle induziert. Um eine optimale Bewurzelung innerhalb von 4 Wochen zu erreichen, werden dabei die kräftigsten Sprosse ausgewählt. Sobald die Bewurzelung eingetreten ist, setzt das endogene rhythmische Wachstum der Eiche wieder ein. Diese bewurzelten Pflanzen werden anschließend in ein halbsteriles Petrischalen-Kultursystem umgesetzt, wo ihre Entwicklung über einen Zeitraum von 10 Wochen untersucht wird. Bei dem System entwickeln sich die Wurzeln zweidimensional und semisteril innerhalb der Petrischale und können mit einer Vorkultur von *P. croceum* direkt beim Anlegen inokuliert werden. Mit dem Kultursystem konnte eine gleichzeitige Beobachtung der Wachstumsschübe des Sprosses sowie eine Gesamterfassung der Wurzelwachstumskinetik (digitale Bildanalyse) vorgenommen werden.
Das endogene rhythmische Wachstum des Sprosses mit Bildung von Wachstumsschüben bleibt unter den optimalen Kulturbedingungen (25°C, 16 h Licht) erhalten. Jeder der sukzessiven Wachstumsschübe wird anhand von vier Entwicklungsstadien beschrieben: A (Endknospe in Ruhephase), B (Endknospe geschwollen), C (Entfaltung der Blätter und Streckungsphase) und D (Ende der Blattentwicklung, Anfang der neuen Ruhephase).

Die Architektur des Wurzelsystems der Stecklinge innerhalb der Petrischalen unterscheidet sich von der stark hierarchisierten Wurzelarchitektur von Eichensämlingen. Dabei führt eine Verringerung der Hierarchie zwischen Haupt- und Seitenwurzeln zu einer intensiveren Seitenwurzelentwicklung mit Produktion entsprechend hoher Mengen an Träger- und Feinwurzeln, so wie sie in der Natur bei älteren Bäumen im Zusammenhang mit der Nährstoffmobilisierung vorkommt. An diesem System wurden die Wachstumskinetiken der unterschiedlichen Wurzeleinheiten untersucht: Trägerwurzeln 1. und 2. Ordnung (TW1 und TW2) bzw. Feinwurzeln 1. und 2. Ordnung (FW1 und FW2). Die wöchentliche Erfassung von Wurzellängen und Wurzelbildung ermöglichte den Nachweis eines rhythmischen Wachstum der Seitenwurzeln.
Ein Vergleich der Wachstumsschübe im Wurzel- und im Sprossbereich zeigte, dass sie zeitlich versetzt stattfinden. Dabei korreliert der Zeitpunkt der maximalen Wachstumsgeschwindigkeit, die die Wurzelwachstumsschübe charakterisieren, mit dem Zeitpunkt des B Stadiums der Wachstumsschübe des Sprosses.

C. Morphologischer Einfluss des EM-Pilzes *Piloderma croceum* auf *in vitro* vermehrte Eichenstecklinge in Dualkulturen und Zusammenhang mit der späten Mykorrhizierung

Die Inokulation der Eichenstecklinge mit dem Ektomykorrhiza-Pilz *Piloderma croceum* im halbsterilen Kultursystem induziert nach 10 Wochen Dualkultur ein sehr starkes Pflanzenwachstum im Vergleich zu Kontrollpflanzen (HERRMANN *et al.* 1998). Das geförderte Wachstum findet einerseits seinen Ausdruck in einer erhöhten Blattfläche bei der nächsten Sprosswachstumseinheit, die sich nach der Inokulation entwickelt. Andererseits ist eine starke Entwicklung des Wurzelsystems zu verzeichnen, die sich insbesondere in einer Zunahme der Gesamtlänge von TW1 und TW2 sowie in einer erhöhten spezifischen Wurzellänge (Verhältnis Wurzelgesamtlänge/Wurzeltrockengewicht) widerspiegelt. Durch die Inokulation mit *P. croceum* wird außerdem die Wurzelhierarchie noch stärker unterdrückt, und die Architektur des Wurzelsystems ist zunehmend mit derjenigen zu vergleichen, die bei älteren Sämlingen zu finden ist. Verzweigungsindices bleiben dahingegen unverändert.
Innerhalb der 10 Wochen Dualkultur tritt die späte Mykorrhizierung erst nach einer langen Vorphase ab der 8. Woche ein. Die o.g. Förderung des Wurzelwachstums durch *P. croceum* findet also vor der Mykorrhizabildung statt. Trotz des frühen Vorhandenenseins von Feinwurzeln werden Mykorrhizen dabei erst nach vollständiger Entfaltung der Blätter (Entwicklungsstadium **D**) gebildet. Obwohl *P. croceum* eine starke Wirkung auf die Wachstumskinetik der Seitenwurzeln ausübt, korreliert auch in seiner Anwesenheit der Zeitpunkt des Sprossent-

wicklungsstadiums **B** mit dem Zeitpunkt der maximalen Wachstumsgeschwindigkeit der Trägerwurzeln.
Durch Zugabe von Aktivkohle oder Indol-3-essigsäure (IES) setzt die Mykorrhizierung bereits ab der 5. Woche ein. Dabei werden durch die Beschleunigung der Mykorrhizierung die morphologischen Effekte von *P. croceum* untedrückt, die sich u.a. bei den TW1 beobachten lassen. Im Anschluss an die frühe Mykorrhizierung ist bei dieser Wurzeleinheit nur noch ein geringes Wachstum festzustellen. Dies ist ein Hinweis darauf, dass die morphologischen Effekte von *P. croceum* spezifisch für die lange Vorphase seiner Mykorrhizierung sind.

D. Molekulare Regulation vor und während der Mykorrhizierung

Nach Optimierung und Charakterisierung wurde das Modell zur Analyse der Regulation der Mykorrhizierung auf molekularer Ebene eingesetzt.
Die molekularbiologischen Untersuchungen wurden an Wurzelmaterial durchgeführt, das sich am Ende der maximalen Wachstumsgeschwindigkeit der TW1 befand, d.h. zum Zeitpunkt, wo sich der Spross zwischen den Entwicklungsstadien **B** und **C** befindet. Zur Erstellung einer subtraktiven cDNA-Bank und für Southern-Blot-Analysen wurden dabei die gesamten neu gebildeten Wurzeln geerntet, während sich die Ernte für Northern-Blot-Analysen auf die neu gebildeten Seitenwurzeln begrenzte.
Für eine Analyse der differentiellen Genexpression in der vormykorrhizalen Phase wurde eine subtraktive cDNA-Bank zwischen Wurzeln in der prämykorrhizalen Phase und einer Mischung von Kontrollwurzeln und Pilz erstellt. Nach Southern-Blot-Analysen von 250 amplifizierten Klonen konnten bisher 65 der differentiell exprimierten Gene sequenziert werden. Etwa die Hälfte der Gene konnte aufgrund einer Homologie-Recherche identifiziert werden.
Eine Northern-Blot-Analyse der Gesamt-RNA von inokulierten versus Kontrollwurzeln zeigte insbesondere eine Hochregulation von Formiat-Dehydrogenase (3,5x) und Metallothionein (3,2x) während der prämykorrhizalen Phase. Solche Northern-Blot-Analysen wurden weiterhin auf Mykorrhizen im aktiven und im seneszenten Zustand erweitert und ergaben erste Hinweise darauf, dass die Expression von Metallothionein-Genen in der Vorphase der Mykorrhizierung spezifisch „hochreguliert“ wird. Metallothioneine sind in der Komplexierung/Detoxifierung von Schwermetallen involviert. Eine hochregulierte Expression von Metallothionein-Genen wurde ebenfalls während der Keimung von Sporen des arbuskulären Mykorrhizapilzes *Gigaspora rosea* durch STOMMEL *et al.* (2001) nachgewiesen.

Schlussfolgerung

Die Feincharakterisierung der Morphogenese im Spross- und Wurzelbereich der Eichenstecklinge ermöglichte die Entwicklung eines *In-vitro*-Modells zur präzisen Untersuchung der Bildung von Ektomykorrhizen zwischen der Steileiche und *P. croceum* und den Nachweis deutlicher morphologischer Effekte in der langen prämykorrhizalen Phase. Auf der Basis dieser Ergebnisse konnten Analysen auf molekularer Ebene gezielt durchgeführt werden. Mit der Erstellung einer subtraktiven cDNA-Bank unter Einsatz der „Suppressive Subtractive Hybridisation" (SSH) Technik wurde in der prämykorrhizalen Phase eine differentielle Expression von 65 Genen gefunden, die sequenziert wurden. Insbesondere wird eine Hochregulation der Expression der Metallothionein- und Formiat-Dehydrogenase-Gene in der prämykorrhizalen Phase nachgewiesen. Die Expression dieser und weiterer Gene in den verschiedenen Phasen der EM-Symbiose wird weiterhin verfolgt. Eine Untersuchung der Einflüsse ausgewählter Faktoren (z.B. Hormone, Zucker, Schwermetalle) wird somit ermöglicht.

Literaturverzeichnis

ALATOU, D.; BARNOLA, P.; LAVARENNE, S.; GENDRAUD, M., 1989: Caractérisation de la croissance rythmique du chêne pédonculé. *Plant Physiology and Biochemistry* 27, 275-280.

HALLÉ, F.; OLDEMAN, R. A. A., 1970: Essai sur l'architecture et la dynamique de croissance des arbres tropicaux. *Monog. Bot. et Biol. Vég.*, Ed. Masson (Paris).

HERRMANN, S.; MUNCH, J.-C.; BUSCOT, F., 1998: A gnotobiotic culture system with oak microcuttings to study specific effects of mycobionts on plant morphology before, and in the early phase of, ectomycorrhiza formation by *Paxillus involutus* and *Piloderma croceum. New Phytologist* 138, 203-212.

KLEBS, G., 1911: Über die Rhythmik in der Entwicklung der Pflanzen. Heidelberger Akademie der Wissenschaften, *Math.-Nat. Kl. Sitzber. Abhandlung* 23, pp 84.

PARMENTIER, C.; BARNOLA, P.; MAILLERD, P.; LAVARENNE, S., 1991: Etude de la croissance rythmique du chêne pédonculé, influence du système racinaire. L'arbre, biologie et développement. C. Edelin Ed. *Naturalia Monspeliensia* n°h.s. 327-343.

REICH, P. B.; TESKEY, R.O.; JOHNSON, P.S.; HINCKLEY, T.M., 1980: Periodic root and shoot growth in oak. *Forest Science* 26, 590-598.

RIEDACKER, A.; DEXHEIMER, J.; TAVAKOL, R.; ALAOUI, H., 1981: Modifications expérimentales de la morphogénèse et des géotropismes dans le système racinaire de jeunes chênes. *Canadian Journal of Botany* 60, 765-778.

RIEDACKER, A.; BELGRAND, M., 1983: Morphogénèse des systèmes racinaires des semis et boutures de chêne pédonculé. *Plant and Soil* 71, 131-146.

SLANKIS, V., 1948: Einfluss von Exudaten von *Boletus variegatus* auf die dichotomische Verzweigung isolierter Kiefernwurzeln. *Physiologia Plantarum* 1, 390-399.

STOMMEL, M.; MANN, P.; FRANKEN, P., 2001: EST-library construction using spore RNA of the arbuscular mycorrhizal fungus *Gigaspora rosea*. *Mycorrhiza* 10, 281-285.

Prozessregulation in der Rhizosphäre
13. Borkheider Seminar zur Ökophysiologie des Wurzelraumes
Hrsg.: W. Merbach, B. W. Hütsch, L. Wittenmayer, J. Augustin
B. G. Teubner Stuttgart, Leipzig, Wiesbaden (2003), S. 20-22

Modelling of root growth of corn (*Zea mays* L.) under field conditions

Rolf KUCHENBUCH
Institut für Primärproduktion und Mikrobielle Ökologie, ZALF, Eberswalder Str. 84, D-15374 Müncheberg

Abstract

Quantitative studies of root development as it depends on climate, soil and plant properties are one prerequisite to derive relations that allow for a more general, possibly predictive approach to root growth and, consequently, root nutrient uptake. Since net maize root length, e.g. amount of roots that can be measured or observed in a soil sample, increases until flowering, detailed information is needed until this stage is entered. The process of dispersion of the total amount of roots formed in soil, i.e. root penetration into soil and root proliferation in soil layers, has to be quantified.

Five field experiments, conducted in three years substantially differing in climate are used to describe root and shoot development on a silt loam soil at one location (West Lafayette, Indiana, USA). Both, total root length development and shoot growth of maize were related to the product of accumulated thermal time and solar radiation for all 5 experiments. Root growth in the topsoil layer was positively related to soil water content as calculated with a mechanistic soil water balance model.

Regression analysis of a long term experiment showed that on a silt loam soil there was a positive relation between root length at flowering in the topsoil layer and the amount of rainfall during the last weeks prior to flowering, e.g. the time of fastest root growth. Root length accumulation at the bottom of the soil profile was positively related to air temperature during the weeks following planting.

Experiments conducted in thin-layer cuvettes filled with a loamy sand soil allowed for a non-destructive and repeated observation of plant root development. The development of a software program enabled quantification of root development on the transparent front of the cuvettes. After penetration of a soil layer root growth was exponential followed by a linear growth phase. Secondary roots were formed at seminal roots approx. 2 days old, and secondary root elongation continued for 6 to 8 days. Only few tertiary roots were observed on scanner images. With decreasing depth the maximum absolute growth rate of roots decreased. Water content of soil layers (SWC) influenced penetration rate of roots into soil as well root proliferation within soil layers. Decreasing SWC increased number of secondary roots and reduced length of secondary roots, the overall effect being a reduction in absolute root growth rate. Increased soil bulk density (SBD) reduced total root length produced in cuvettes. The main reason was reduction of number of secondary roots formed. Seminal root development was not strongly influenced by soil bulk density. With decreasing depth the effect of SBD on the formation of lateral roots increased strongly.

It has been shown that during vegetative growth of the corn plant shoot growth and root abundance are interdependent, and a maximum root-shoot ratio (RSR) can be derived. Root penetration into soil and root proliferation during the exponential and linear growth phase in soil layers (root dispersion) can be described sufficiently exact by regression equations that take into account air temperature and solar radiation. A maximum absolute penetration rate (APR) and growth rate (AGR) can be derived. AGR and APR are influenced by distance from soil surface, soil water content and soil bulk density; these relations can be expressed mathematically.

On basis of the results obtained a one-dimensional mathematical model is developed for a layered soil using STELLA software. Input data are daily values for minimum and maximum air temperature, and solar radiation. The soil profile is subdivided in up to 10 layers with a minimum depth of 5 cm, and each layer is characterized by one soil bulk density. Daily values of soil water contents of soil layers are calculated using an external model of soil water balance. From these data the model calculates shoot biomass accumulation, root abundance from shoot growth, root penetration into soil, and root proliferation in soil layers. Maximum absolute root growth rate and maximum penetration rate are altered with soil depth, with changes of soil water content for so il layers over time, and as soil bulk density differs for soil layers.

The following rules were implemented:

1. Root length produced per day cannot exceed the value resulting from shoot growth and RSR, e.g. maximum root growth is strictly dependent upon shoot biomass accumulation. Weather conditions alter shoot growth and hence alter root abundance.
2. Root proliferation in any one-soil layer cannot start before the root has penetrated to that depth of the soil profile. The sum of root length produced in all soil layers with roots cannot exceed maximum root abundance.
3. In each soil layer root growth starts with an exponential phase followed by a linear phase later on. One maximum AGR of the root applies to all soil layers.

As could be expected for the data set used for model development, calculations agreed well with data of shoot growth and root abundance. Root growth in soil layers was best reflected for the 0-15 and 15-30 cm layer, and deviations between measured and predicted values were highest for the deepest soil layers.

It is concluded that the root growth model reflects the major processes that determine root abundance and dispersion for the weather and soil conditions that are characteristic for the location. Inclusion of other soil types and climates may extend the applicability of the model to other environmental conditions in the future.

The model presented here is simplistic in a number of aspects. The following list gives some keywords and is not final:

- The root growth model is one-dimensional and roots are evenly distributed in the soil volume; clustering of roots and possible effects on nutrient and water acquisition are not represented.
- The soil related algorithms incorporated in the root model are site specific; the influence of soil types differing in pore space, pore diameter and pore continuity are not represented.
- The possible influence of adverse soil chemical conditions (local response) is not incorporated; salinity effects and aluminum toxicity are not represented.
- Turnover processes in soil that are mediated by microorganisms are not included; root decay, nitrogen liberalization are not represented.
- Processes that influence nutrient availability other than physico-chemical are not included, e.g. effects of root exudation and mycorrhiza infection are not represented.
- Different binding forms of ions in soil are not implemented, e.g. non-exchangeable potassium that has been shown to be available to plants, is not represented.
- The influence of heat, cold and drought stresses on shoot (and root) growth are not represented.

The list shows that model calculations cannot be expected to agree well with observations if one or more of these factors are dominating over the processes that are at present incorporated into the model. However, the degree of agreement may be used to decide whether or not additional processes have to be included. The presented, simplistic model can be used as a starting point.

Prozessregulation in der Rhizosphäre
13. Borkheider Seminar zur Ökophysiologie des Wurzelraumes
Hrsg.: W. Merbach, B. W. Hütsch, L. Wittenmayer, J. Augustin
B. G. Teubner Stuttgart, Leipzig, Wiesbaden (2003), S. 23-31.

Significance of root growth and activity for genotypic differences in nitrogen efficiency of rapeseed (*Brassica napus* L.): I. Estimation of root growth and nitrate-N uptake under field conditions

Mahmoud KAMH[1], Franz WIESLER[2], and Walter J. HORST[3]
[1] Soil and Water Science Dept. Faculty of Agriculture, University of Alexandria, El-Shatby, Alexandria, Egypt.
[2] LUFA Speyer, Obere Langgasse 40, D-67346 Speyer, Germany.
[3] Institute of Plant Nutrition, University of Hannover, Herrenhäuser Str. 2, D-30419 Hannover, Germany.

Abstract

Nitrate-N uptake from soil depends on root growth and uptake activity. However, under field conditions nitrate-N uptake is difficult to estimate due to different losses which can be hardly controlled. We modified a current method (mesh bag method) to estimate nitrate-N uptake activity along with the root production under field conditions. PVC tubes having windows on the upper side and an anion exchange resin on the bottom were installed into the root zone of oilseed rape at depths of 0–30 and 30–60 cm, then filled with root-free soil which was amended with and without 90 mg N per kg soil. After the installing period, the resin and soil subsamples were analysed for NO_3^--N and in the remainder soil root growth was determined. Nitrate-N depletion was estimated from differences between NO_3^--N in the compartments and NO_3^--N in control compartments installed under the same conditions. Nitrate leached from the compartments was quantified from the resin and has been taken into consideration in our calculation. N application into the compartments resulted only in a small increase of root growth but allowed estimation of N uptake. N depleted from the compartments significantly correlated with root-length density. However, N application in the field as well as root-length density in the compartments (> 5 cm cm^{-3}) affected the results. The method is suitable for estimating relative nitrate uptake activity and root production (comparing plants growing with different N fertilization rates or comparing cultivars) under field conditions.

Introduction

Nitrogen uptake-efficiency depends on the extent and effectiveness of the root system (JACKSON et al. 1986) so that cultivars with deeper root system and higher root-length densities deplete more nitrate-N from the soil (WIESLER and HORST 1994, OIKEH *et al.* 1999). But there is often a weak correlation between root-length density and N uptake. Nitrate-N uptake by plant roots depends not only on root-length density but also on root activity (REIDENBACH and HORST 1997; GAO *et al.* 1998).

The root system is dynamic and its net growth is a balance between root production and root mortality. Root production depends on a number of factors, e.g. carbohydrate supply from the shoot (GAO *et al.* 1998, THALER and PAGES 1996 und 1998) and nutrient supply in the growth medium (BLOOM 1997). Increasing assimilate supply during the reproductive growth phase improves tolerance to N stress by maize (BÄNZIGER *et al.* 2002) and by rapeseed (DECCER *et al.* 2000, WIESLER *et al.* 2001). Root production during a given period of time represents the ability of the plant to exploit the soil and can be estimated by the mesh bag (in-growth core) method (STEEN, 1984, STEINGROBE *et al.* 2000). Although the effect of nutrient supply and growth stage on root production and morphology have been frequently studied (HAIRIAH et al. 1991, PRICE and HENDRICK 1998, STEINGROBE *et al.* 2000, STEINGROBE 2001) there is no work on direct assessment of nutrient uptake activity under field conditions.

Nitrogen is taken up by plants in the form of NH_4^+ and NO_3^-. Nitrate-N is very mobile in the soil and is liable to different loss pathways, namely leaching and denitrification. Therefore, estimating N uptake under field condition is difficult, and there is still no method available for such an estimation. The aim of this study was to estimate root production and nitrate-N uptake activity of oilseed rape grown with limited and non-limited N supply under field conditions.

Materials and methods

Oilseed rape (*Brassica napus* L.) was sown (83 seeds m^{-2}) in August 2001 in a field near Göttingen, southern Lower Saxony. The soil was loamy with 1.3 mg total nitrogen g^{-1}, and 12 mg total carbon g^{-1}, and 13.9 kg N ha^{-1} as available N_{min} (0– 90 cm). Rape was grown on plots sized 11.25 m^2 and containing 6 plant rows. The plots received 0 (N0) or 227 (N2) kg N ha^{-1} as ammonium-nitrate fertilizer. The experiment was laid out in a split plot design with four replicates. In February 2002, holes were drilled into the soil between plant-rows with an angle of 45° and 5 cm diameter for 2 different soil depths (0-30 and 0-60 cm as vertical depth). Immediately after drilling the holes, PVC tubes with a diameter of 5 cm were inserted into the holes. At the beginning of shooting (March), beginning of

flowering (May), and at seed filling (June), the PVC tubes were substituted by compartments.
The compartment is also a PVC tube of the same diameter but having a window of 3.4 cm width and 42.5 cm length at the upper side. The window was positioned 5 cm above the tube's bottom, while the tube's length depended on the soil depth under study (0-30 or 30-60 cm depth). A ring (4.2 cm diameter and 1 cm length), covered with nylon screen 40 μm on both opining sides, and containing 12 g of an anion exchange-resin (Dowex 1x8, 20-50 mesh, Cl^- form), was fixed on the bottom of the compartment which had a hole of 4 mm diameter. Above the resin ring, a sand filter of the same diameter and 3 cm length was fixed using a rubber to close any space between the filter as well as the resin ring and the compartment. This construction was used to quantify the amount of nitrate which could be leached from the compartment.
After inserting into the holes, the compartments were filled with root-free soil from the same field at a bulk density of 1.35 g cm^{-3}. The soil was collected from the field of the N0 treatment at a depth below 30 cm. After air drying and passing through a 4 mm sieve, N_{min} of the soil was extracted by 1 mM KCl and the extract was measured using an autoanalyser. One day before filling the compartments, $Ca(NO_3)_2$ was added to 1 kg soil samples if necessary to increase the N_{min} content to 100 mg kg^{-1} and the moisture content to 13 % (50 % of the water holding capacity), then kept at 4°C till the next day. Soil samples were transported to the field in cooling bags. After filling, the openings (above-ground) of the compartments were closed with plastic bags. Four compartments were installed per plot, two compartments in which the soil was treated with N, one without N (see above), and one as a control. The control is a compartment in which the window was covered by a 0.65 μm membrane in order to exclude root growth into the compartment but allowing all other kind of equilibration with the bulk soil in the field. After 21 days (shooting) and 16 days (flowering and seed filling), the compartments were pulled out. The resin ring and the soil were kept at 4°C and -20 °C, respectively, until analysis.
The resin was subjected to 5 consecutive extraction using 5% KCl solution followed by washing the resin with KCl solution to obtain a total volume of 100 ml. Nitrate-N in the extraction solution was then measured by an autoanalyser. Subsamples of 200 g soil were taken and used for measuring N_{min} and the moisture content. Roots were washed from the remainder soil samples and the root length was estimated by RHIZO image analysis. Nitrate-N depletion from compartments were calculated by subtracting (N_{min} + N_{resin}, of compartments) from (N_{min} + N_{resin}, of control compartments).
In order to test the capacity of the resin for NO_3^--N adsorption, 5 compartments were placed on a holder at 45° in a green house. The windows were covered with

nylon nets, and the compartments were filled with treated soil to the above bulk density. Compartments were subjected to excessive leaching. Then the resin and the soil were taken for NO_3^--N analysis.

Results

Excessive leaching of NO_3^--N from the compartments (simulating a very heavy rain storm) showed that the amount of resin used can bind about half of the NO_3^- (53 mg kg^{-1}) applied to +N compartments (Tab. 1).

Tab. 1. NO_3^--N adsorbed by resin in the compartment after excessive leaching (500 ml effluent)

Initial soil N	Residual soil N	Resin N	Leachate N	N balance
		mg N $compartment^{-1}$		
106	17.8	53.4	28.8	- 6.0

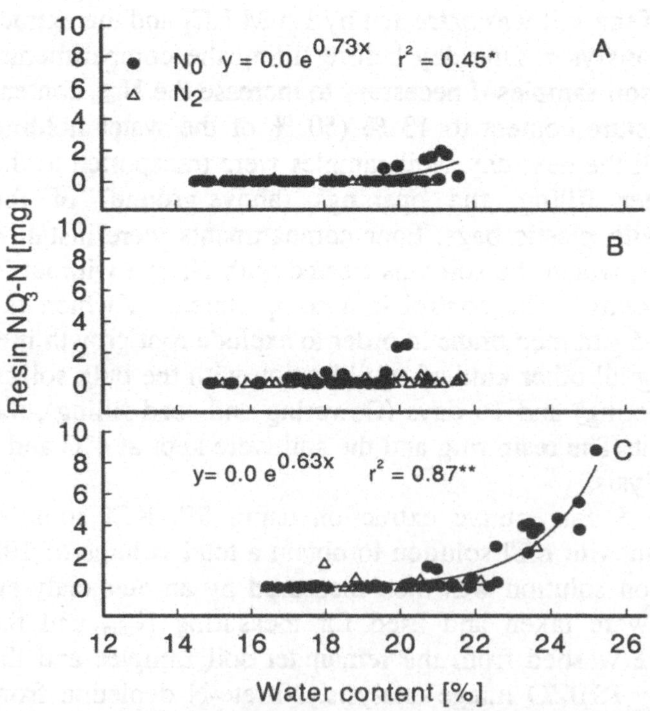

Fig. 1. Relationships between soil water content and the amount of NO_3^--N leached from compartments installed in an oilseed-rape field for 21 days at shooting (A), for 16 days both at flowering (B), and at seed filling (C).

Under field conditions, the maximum amount of NO_3^--N leached from the compartments did not exceed 10% (Fig. 1) and occurred especially in compartments in-stalled in the upper 0–30 cm soil.

During shooting and flowering, the moisture content in the compartments (measured after the installing period) ranged between 16 and 22%, depending on soil depth and plots, with maximum leaching of 3 mg NO_3^--N. During the seed-filling period, both moisture content and NO_3^--N leaching were higher, due to much higher precipitation.

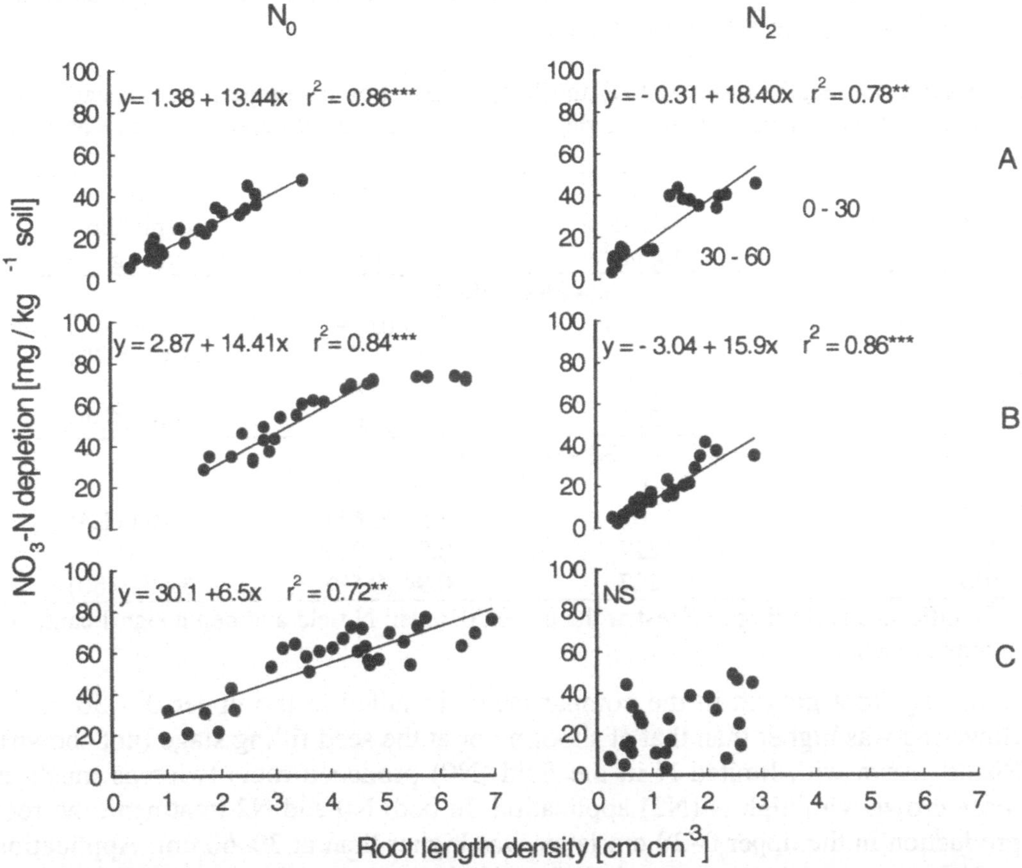

Fig. 2.. Relationships between root-length density and NO_3^--N depletion from compartments installed in an oilseed-rape field for 21 days at shooting (A) and for 16 days both at flowering (B) and at seed filling (C).

The depletion of nitrate-N in +N compartments was significantly correlated with root-length density (RLD) at shooting, flowering, and seed filling except for N2 (Fig. 2 A, B, and C, respectively). At shooting, the data obtained from compartments installed in the surface soil layer of N2 plots did not appear to be correlated with RLD (Fig. 2A). This might be due to N application in the field one week before installing the compartments. At flowering and seed filling, the RLD was much higher than that at shooting, especially in the N0 treatment. At RLD higher than 5 cm cm^{-3}, the measured N_{min} values were close to zero. This restricted the detection of the actual depletion capacity of the roots in such compartments (Fig. 2B and C).

Root-length density and nitrate-N depletion as averaged over two rape cultivars during flowering are presented in Table 2.

Tab. 2. Root-length density (RLD) and NO_3^--N depletion in compartments installed in an oilseed-rape field at two soil depths for 16 days at flowering (means of two cultivars)

N compartments mg kg^{-1}	N field kg ha^{-1}	RLD cm cm^{-3}	N depletion mg kg^{-1}
	0 – 30 cm depth		
0	0	4.10 (0.74)	--
100	0	4.98 (1.07)	68.40 (6.40)
0	227	1.46 (0.82)	--
100	227	1.23 (0.63)	20.96 (10.90)
	30 – 60 cm depth		
0	0	2.30 (0.47)	--
100	0	2.58 (0.63)	43.10 (11.4)
0	227	0.91 (0.31)	--
100	227	0.94 (0.50)	15.60 (9.80)

Significance according to F test at the $p < 0.001$level: N field and depth significant, N compartment ns

Generally, root growth in the compartments installed in the upper 0 – 30 cm at flowering was higher than that at shooting or at the seed filling stage (not shown). Plants grown with limited N in the field (N0) produced roots twice as much as those grown with high N (N2) application. In both N0 and N2 treatment, the root production in the upper 0–30 cm layer was higher than at 30–60 cm. Application of N in the compartments resulted in a slight increase of root production. This increase was more pronounced in the N0 treatment, especially in the surface soil-layer.

Without N application in the compartments (-N compartments), where values of N_{min} were always close to zero, no depletion of nitrate N could be calculated. In +N compartments, more N was depleted in the upper 0–30 cm compared to subsoil and was higher in N0 compared to N2.

Discussion

There is no single method of root measurements applicable for all purposes and the method of choice depends on many factors (VOGT et al. 1998). Our major interest was to estimate the nitrate-N uptake directly along with root production of oilseed rape under field conditions. Our method was based on the in-growth core method (STEEN 1984, STEINGROBE et al. 2000) with modifications allowing the estimation of N uptake.

Growth of roots into the root-free soil in the compartment during the installing period (2-3 weeks) was used to estimate the root production. As reported for the in-growth core method, variation of nitrate supply, soil density, and moisture content of the repacked soil compared to the bulk soil outside the core can influence root growth (VOGT et al. 1998, STEINGROBE et al. 2000). The compartments were filled with the soil at a bulk density nearly similar to that in the field (1.35 g cm^{-3}). Bulk density of repacked soil in the range of 1.0 to 1.5 gm cm^{-3} did not significantly affect root development in mesh bags (HAIRIAH et al. 1991). The initial moisture content of the repacked soil was 13% which was equilibrated later with the moisture content in the field (as measured after 2 days in both the bulk soil and test compartments, not shown). The soil moisture contents in the compartments after the installing period were nearly similar to that in the field during shooting and flowering. During seed filling, a 1-3% increase of the moisture content was measured in the compartments compared to the field, due to heavy rain which might have drained from the compartments not quickly enough.

Although not significant, localization of nitrogen in the compartments tended to enhance root growth by 3–20%, depending on field-N treatment, growth stage, and soil depth which may reflect differences between soil N in the compartments and in the field. The enhancement was more pronounced at flowering in the N0 field (Tab. 2), especially in the upper 0-30 cm soil. Many plant species respond to localized N supply by an increase of root growth and branching (ROBINSON, 1994, DREW et al. 1973). In agreement with EGHBALL and MARANVILLE (1993) plants grown with non-limited N in the field produced less roots than those grown with limited N, independent of N application in the compartments (Tab. 2). This indicates that root production depends mainly on the internal N status of the plant.

The present method allowed the assessment of nitrate-N uptake by the newly produced roots. Depending on N_{min} in the compartments compared to that in the control, the amount of nitrate-N depleted from the compartments was significantly correlated with the root-length density. In this context the amount of nitrate leached from the compartments was collected on the resin and was considered in our calculations. High root length density in the compartments as well as N fertilization in the field can affect the result as reflected by deviations of some compartments from the linear relationship (Fig. 2). Such factors should not be defined as limiting factors since they could be eliminated. Denitrification of nitrate was shown to respond rapidly to soil wetting, due to partial decrease in O_2 availability (DAVIDSON 1992), and stimulated by C addition (MCKENNEY et al. 2001). Since after the observation period soil moisture contents in the control, non-rooted compartments, were in the same range as those of the rooted compartments we assume that the denitrification occurred at a similar magnitude in both compartments. However, carbon supply from the root may affect the microbial activity and hence the rate of denitrification. An adjustment of the method may require the direct estimation of nitrate losses through denitrification using ^{15}N. However such an optimization is difficult under field conditions.

References

BÄNZIGER, M.; EDMEADES, G. O.; LAFITTE, H. R., 2002: Physiological mechanisms contributing to the increased N stress tolerance of tropical maize selected for drought tolerance. *Field Crop Res.* 75, 223-233.

BLOOM, A. J., 1997: Interaction between inorganic nitrogen nutrition and root development. *Z. Pflanzenernähr. Bodenk.* 160, 253-259.

DAVIDSON, E. A., 1992: Sources of nitric oxide and nitrous oxide following wetting of dry soil. *Soil Sci. Soc. Am. J.* 56, 95-102.

DRECCER, M. F.; SCHAPENDONK, A. H. C. M.; SLAFER, G. A.; RBBINGE, R., 2000: Comparative response of wheat and oilseed rape to nitrogen supply: absorption and utilization efficiency of radiation and nitrogen during the reproductive stages determining yield. *Plant Soil* 220, 189-205.

DREW, M. C.; SAKER, L. R.; ASHLEY, T. W., 1973: Nutrient supply and the growth of the seminal root in barley: I. The effect of nitrate concentration on the growth of axes and laterals. *J. Exp. Bot.* 24, 1189-1202.

EGHBALL, B.; MARANVILLE, J. W., 1993: Root development and nitrogen influx of corn genotypes grown under combined drought and nitrogen stresses. *Agron. J.* 85, 147-152.

GAO, S.; PAN, W. L.; KOENIG. R. T., 1998: Integrated root system age in relation to plant nutrient uptake activity. *Agron. J.* 90, 505-510.

HAIRIAH, K.; VAN NOORDWIJK, M.; SETIJONO, S., 1991: Tolerance to acid soil conditions of the velvet beans *Mucuna pruriens* var. *utilis* and *M. deeriniana*. I. root development. *Plant Soil* 134, 95-105.

JACKSON, W. A.; PAN, W. L.; MOLL, R. H.; KAMPRATH, E. J., 1986: Uptake, translocation, and reduction of nitrate. In: Neyra, c.a. (Ed), Biochemical Basis of Plant Breeding. Nitrogen Metabolism, Vol 2. CRC press, Boca Raton, FL, pp. 95-98.

McKenney, D. J.; Drury, C. F.; Wang, S. W., 2001: Effect of oxygen on denitrification inhibition, repression, and derepression in soil columns. *Soil Sci. Soc. Am. J.* 65, 126-132.

Oikeh, S. O.; Kling, J. G.; Horst, W. J.; Chude, V. O.; Carsky, R. J., 1999: Growth and distribution of maize roots under nitrogen fertilization in plinthite soil. *Field Crop Res.* 62, 1-13.

Price, J. S.; Hendrick, R. L., 1998: Fine root length production, mortality and standing root crop dynamics in an intensively managed sweetgum (*Liquidabar styraciflua* L.) coppice. *Plant Soil* 205, 193-201.

Reidenbach, G.; Horst, W. J., 1997: Nitrate-uptake capacity of different root zones of *Zea mays* (L.) in vitro and in situ. *Plant Soil* 196, 295-300.

Robinson, D., 1994: The responses of plants to non-uniform supplies of nutrients. *New Phytol.* 127, 635-674.

Steingrobe, B.; Schmid, H.; Classen, N., 2000: The use of the ingrowth core method for measuring root production of arable crops-influence of soil conditions inside the ingrowth core root growth. *J. Plant Nutr. Soil Sci.* 163, 617-622.

Steingrobe, B., 2001: Root renewal of sugar beet as a mechanism of P uptake efficiency. *Plant Nutr. Soil Sci.* 164, 533-539.

Steen, E., 1984: Variation of root growth in a grass ley studied with a mesh bag technique. *Swedish J. Agric. Res.* 15, 25-30.

Thaler, P.; Pages, L., 1996: Root apical diameter and root elongation rate of rubber seedling (*Hevea brasiliensis*) show parallel responses to photoassimilate availability. *Physiol. Plant.* 97, 365-371.

Thaler, P.; Pages, L., 1998: Modelling the influence of assimilate availability on root growth and architecture. *Plant Soil* 201, 307-320.

Vogt, K.A.; Vogt, D.J.; Bloomfield, J., 1998: Analysis of some direct and indirect methods for estimating root biomass and production of forests at an ecosystem level. *Plant Soil* 200, 71-89.

Wiesler, F.; Horst, W. J., 1994: Root growth and nitrate utilization of maize cultivars under field conditions. *Plant Soil* 163, 267-277.

Wiesler, F.; Behrens, T.; Horst, W.J., 2001:. Nitrogen efficiency of contrasting rape ideotypes. In: Horst et al. (Eds.) Plant Nutrition - Food Security and Sustainability of Agro-Ecosystems. Kluwer Academic Publishers, Dordrecht Boston London. pp 60-61.

Prozessregulation in der Rhizosphäre
13. Borkheider Seminar zur Ökophysiologie des Wurzelraumes
Hrsg.: W. Merbach, B. W. Hütsch, L. Wittenmayer, J. Augustin
B. G. Teubner Stuttgart, Leipzig, Wiesbaden (2003), S. 32-40

Significance of root growth and activity for genotypic differences in nitrogen efficiency of rape (*Brassica napus* L.):
II. Root production and nitrate-N uptake by two cultivars

Mahmoud KAMH[1], Franz WIESLER[2], Abdullah ULAS[3], and Walter J. HORST[3]
[1] Soil and Water Science Dept. Faculty of Agriculture, University of Alexandria, El-Shatby, Alexandria, Egypt.
[2] LUFA Speyer, Obere Langgasse 40, D-67346 Speyer, Germany.
[3] Institute of Plant Nutrition, University of Hannover, Herrenhäuser Str. 2, D-30419 Hannover, Germany.

Abstract

Pervious work demonstrated a suitable method (compartment method) for estimating root-growth and nitrate-N uptake activity in the field. In this work, the compartments were installed in the plots of two contrasting oilseed rape cultivars (N-efficient 'Apex' and N-inefficient ''Capitol'') fertilized with 0 (N0) and 227 (N2) kg N ha^{-1}. The new developed roots and nitrate-N depletion were estimated in the compartments after the installing period (21 days at shooting and 16 days both at flowering and grain-filling stages).

Suboptimum N application to the crop (particularly N0) reduced total biomass and seed yield formation substantially (24% and 38% for 'Apex' and 'Capitol', respectively). At the shooting stage, there were no differences in root production and N uptake of both cultivars between N0 and N2. But at flowering and seed filling stages higher root production and accordingly higher N depletion was observed at N0 compared to N2. Towards later growth stages the new developed roots were characterised by reduction of root diameter and a shift towards the deeper soil layer (30-60 cm). At low N but not at high N supply, the N-efficient cultivar 'Apex' exhibited higher root-growth activity and accordingly depleted nitrate-N more effectively than the N-inefficient cultivar 'Capitol' especially during the reproductive growth phase. The calculated nitrate-N uptake rate per unit root length was maximum at flowering but showed no differences between the two cultivars. This indicated that the higher N uptake efficiency of 'Apex' was due to higher root-growth activity rather than higher uptake per unit of root length.

Introduction

The improvement of soil and fertilizer N management and the breeding of N-efficient genotypes may contribute to optimizing yields using less N fertilizer. This will reduce environmental risk in high input agriculture and sustain higher yields in low input agriculture.

Previous studies have shown genotypic differences in N efficiency among oilseed rape genotypes (BEHRENS 2002, GRAMI and LA GROIX 1977, WIESLER *et al.* 2001a). At low N supply, differences between 12 cultivars in their N efficiency (seed yield per N supply) could not be explained by N-utilization efficiency (WIESLER *et al.* 2001b). Based on the seed yield, the cv. 'Apex' was classified as N-efficient and 'Capitol' as inefficient. The calculated N uptake between beginning of flowering and maturity indicated higher N uptake activity of 'Apex' than 'Capitol', suggesting that yield formation and nitrogen use efficiency were closely related to N-uptake efficiency during the reproductive growth phase (WIESLER *et al.* 2001b, HORST et al. 2002).

In our previous work (KAMH *et al.* this book), we demonstrated a suitable method to estimate nitrate-N uptake activity along with root production under field conditions. The aim of this work was to study root production and nitrate-N uptake of two contrasting rape cultivars, 'Apex' (N-efficient) and 'Capitol' (N-inefficient) at different growth stages.

Materials and methods

Two cultivars of oilseed rape (Brassica napus L.) were sown in August 2001 in a field near Göttingen, southern lower Saxony. The cultivars were N-efficient 'Apex' and N-inefficient 'Capitol'. The experiment was laid out in a split plot design in which plots sized 11.25 m^2 received 0 (N0) or 227 (N2) kg N ha^{-1} as ammonium nitrate fertilizer in addition to the N_{min} in the soil (13 kg N ha^{-1}). In February 2002, holes were drilled into the soil between plant-rows with an angle of 45° and 5 cm diameter for 2 different soil depths (0-30 and 0-60 cm). PVC tubes of 5 cm outer-diameter were inserted into the holes. At shooting (March) , flowering (May), and at seed filling (June), the PVC tubes were substituted by compartments which were then filled with root-free soil amended with and without 90 mg N kg^{-1}. For more details about the compartments, procedures, and analysis see KAMH *et al.* (this book). The compartments were installed for 21 days at shooting and for 16 days both at flowering and at seed filling stages. The new developed roots, N_{min}, and the leached N were quantified in the compartments and used to calculate the NO_3^--N uptake.

Results

Biomass formation at harvest and seed yields were lower at low N0 compared to high N supply (N2) (Tab. 1). Seed yield of cv. 'Apex' was higher than that of 'Capitol' at low but not at high N supply.

Tab. 1. Seed yield and total biomass of two oilseed rape cultivars grown in the field in 2002.

N application	0 kg N ha^{-1}		227 kg N ha^{-1}	
Cultivar	'Apex'	'Capitol'	'Apex'	'Capitol'
Seed yield [t ha^{-1}]	3.27 a	2.77 b	4.28 a	4.43 a
Total biomass [t ha^{-1}]	7.71 a	7.52 a	9.50 a	10.75 a

Means within N treatment and yield followed by the same letter don't differ significantly at $p < 0.05$ (Tukey-test).

Root-length densities in the compartments after 21 days at shooting, and 16 days at flowering and seed filling stages were independent of N application in the compartments (Fig. 1).

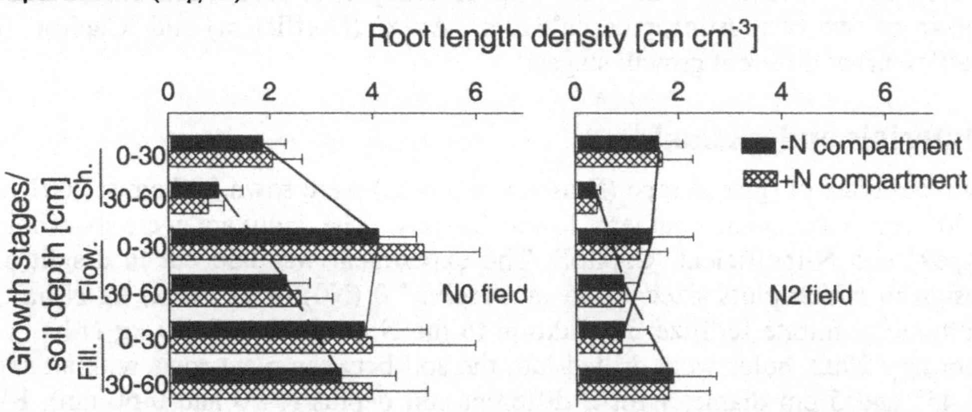

Fig. 1. Root growth (average of two cultivars) at two profile depths and at three growth stages (Sh. = shooting, Flow. = flowering, Fill. = grain filling) in the compartments installed in an oilseed rape field as influenced by the N supply to the crop in the field (NF) and in the compartment (NC). F-test: ns: NC, NF x NC, growth stage x NC; significant at p< 0,01: NF, at p< 0,001: depth, growth stages

However, N-limited rape plants at N0 showed much greater root-growth activity in the compartments especially at the flowering and seed filling stages. In both N0 and N2, root production in the subsoil layer (30-60 cm) increased with growth stages till seed filling. At this growth stage root-length densities were equal (N0)

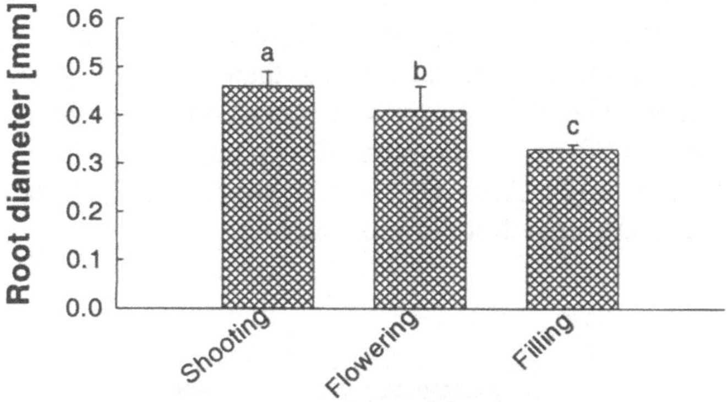

Fig. 2. Average diameter of the roots grown in the compartments at three growth stages. Means with the same letters are not significantly different at the P > 0.05 level (Tukey-test).

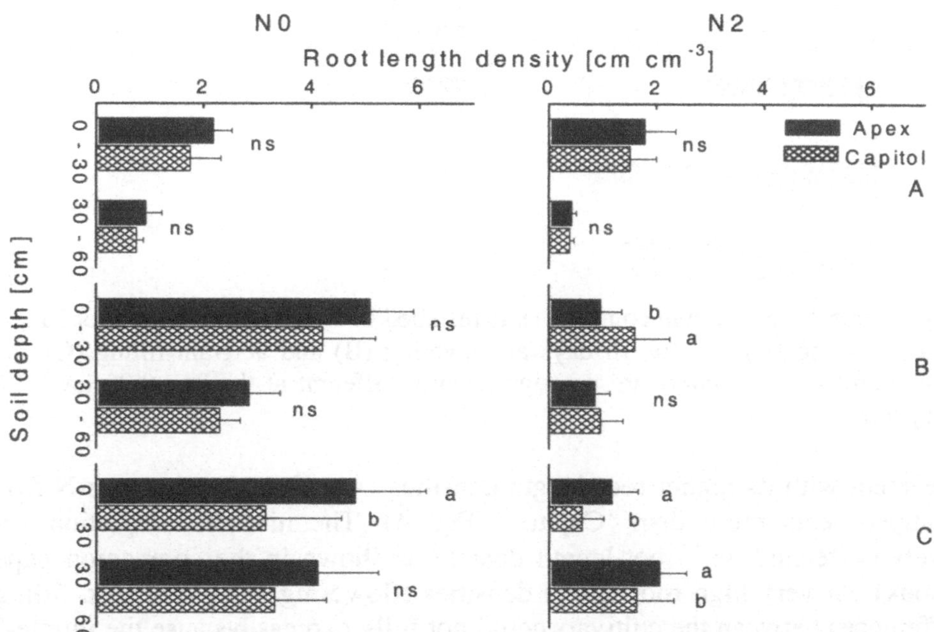

Fig. 3. Root growth in the compartments installed in an oilseed rape field for 21 days at shooting (A), and for 16 days at flowering (B) and at grain filling (C). Means with the same letters are not significantly different at the P < 0.05 level (Tukey-test).

or greater (N2) than in the surface soil. The diameter of the roots growing into the compartments decreased from shooting to seed filling (Fig. 2). Since the root

diameter did not significantly differ between N0 and N2 plants as well as between cultivars means over N supplies and cultivars are presented.
The two cultivars differed in root production at limited N supply (N0) (Fig. 3). Generally, 'Apex' was able to produce more new roots than 'Capitol' at the flowering and seed filling stages. The differences were most pronounced and significant in the upper soil layer (0-30 cm) at the seed filling stage when the root length densities were particularly high. Also at high N supply (N2) 'Apex' produced more roots at the seedfilling stage but not at flowering.

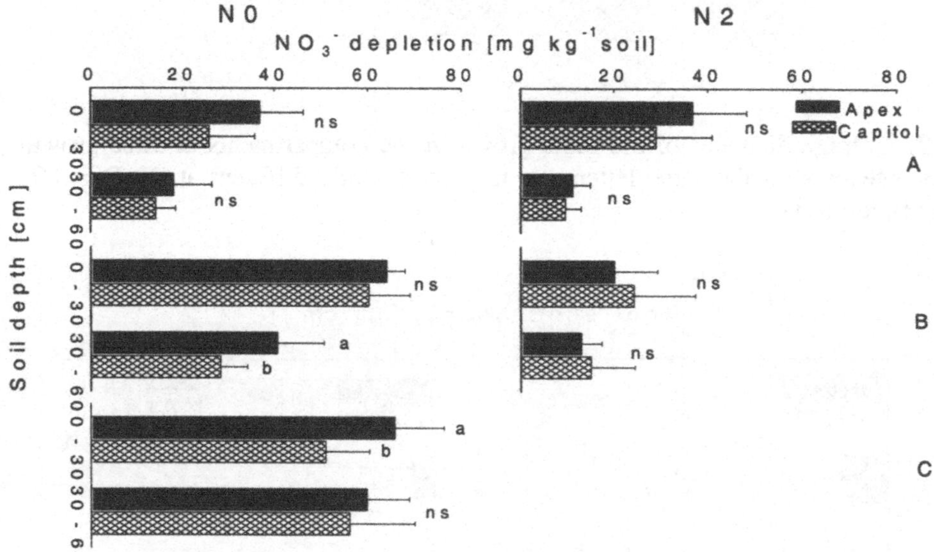

Fig. 4. Nitrate-N depletion in compartments installed in an oilseed rape field for 21 days at shooting (A) and for 16 days at flowering (B) and at grain filling (C). Means with the same letters are not significantly different at the $P < 0.05$ level (Tukey-test).

In agreement with its higher root-length densities 'Apex' depleted nitrate-N from the compartments more than 'Capitol' (Fig. 4). The nitrate-N depletion was positively correlated with root-length density as shown in the companion paper (this book). At very high root length densities (flowering, 0-30 cm; seed filling) the differences between the cultivars could not fully express because the nitrate-N was nearly fully depleted during the measuring period. At high N supply (N2) the compartments were less depleted in nitrate-N and no consistent differences existed between the cultivars.
From the depletion of nitrate-N, uptake rate per unit of root length in the compartments could be calculated for each growth stage.
No differences existed between 'Apex' and 'Capitol' (Tab. 2). Nitrogen-limited

(N0) plants showed the highest nitrate-N uptake rate at flowering. Only at shooting N2 plants showed a significantly higher uptake rate than N0 plants. At the other growth stages the uptake rate tended to be lower.

Tab. 2. Nitrate-N uptake rate of two oilseed rape cultivars grown in the field in 2002 at 0 (N0) and 227 (N2) kg ha^{-1} N supply.

Cultivars	NO_3-N uptake [pmol cm^{-1} s^{-1}]					
	Shooting		Flowering		Grain filling	
	N0	N2	N0	N2	N0	N2
'Apex'	0.98	1.43	1.23	1.29	1.16	1.17
	(0.16)	(0.38)	(0.20)	(0.18)	(0.25)	(0.46)
'Capitol'	1.03	1.40	1.27	1.16	1.21	1.09
	(0.21)	(0.43	(0.16)	(0.28)	(0.16)	(0.34)

F-test: ns: Cutivars;
significant at the $p < 0,01$: N, at the $p < 0,001$: stage x N.

Discussion

Morphogenetic effects of N supply on root growth are well documented (BLOOM 1997, FORDE and LORENZO 2001, SATTELMACHER and THOMS 1995). Generally, high, supraoptimal N supply leads to a reduction in root length/rooting depth but to an enhanced branching of the roots (ANDERSON 1987, COMFORT *et al.* 1988, DURIEUX *et al.* 1994). Particularly a localised N supply stimulates branching in the zone of N supply (DREW et al. 1973, GRANATO and RAPER 1989, STEINGROBE *et al.* 2000). Surprisingly, in our experiments N application to the compartments did not affect root growth into the compartments of N unlimited (N2) as well as at N limited plants (N0) (Fig. 1). At shooting this might have been due to the fact that at this growth stage, the N supply of the soil was still high enough at both fertilizer levels to meet the young plants' N demand. At later growth stages, root-length densities were about twice as high at N0 compared to N2. This shows that the N status of the plants more than the N supply to the roots affected root growth under the prevailing conditions. In absence of stress factors in the root zone carbohydrate supply to the roots is a main factor affecting root growth (THALER and PAGES 1998). It appears that N deficiency results in increased carbohydrate sink-strength of the roots compared to other sinks, greater allocation of assimilates to the roots and thus increased new root growth (MERRILL *et al.* 2002, ZAGAL 1994). At high N supply, at the reproductive growth stages most of the assimilates will be directed to reproductive plant parts such as florescences and pods. The need for new root growth is less since the N requirements for rape seeds can mostly be met by retranslocation from vegetative plant organs (BEHRENS 2002).

Root production of oilseed rape in the 0–30 cm soil layer is greatest at flowering (Fig. 1, BARRACLOUGH 1989, KAPPEN *et al.* 2000). The reduction at seed filling may reflect less root activity in this root zone at this growth stage due to root decay as proposed by WIESLER and HORST (1993) and BARRACLOUGH (1989). Higher root mortality within the compartments is unlikely because the growth period of 16 days was less than the expected life span of a single root (STEINGROBE 2000). The reduction in the top soil was associated with increased root length in the deeper soil layer reflecting increasing root growth activity in the subsoil at later developmental stages thus maintaining N and water uptake as has been observed by other authors (OIKEH *et al.* 1999).

Newly developed roots are of particular importance for nutrient uptake (GAO *et al.* 1998). This is supported by the close relationship between root growth in the compartments and nitrate-N depletion which is indicative of nitrate uptake (KAMH *et al.* this book). The results show differences between the two cultivars in their root production and nitrate depletion, especially at the flowering and seed filling stages at limiting N supply (N0) (Figs. 3, 4). These differences are in agreement with the results reported by HORST *et al.* (2002), who, using the mini-rhizotron method, demonstrated higher root-growth activity for 'Apex' than for 'Capitol' in the field.

The calculated nitrate-N uptake rates (Tab.2) are at the lower margin of the uptake rates estimated by other authors for other crops (REIDENBACH and HORST 1997) and do not support the view that other losses from the compartments than uptake were of major importance. The two cultivars did not differ in their nitrate-N uptake per unit root length at all growth stages. This indicates that roots took up nitrate near I_{max} and, therefore, uptake depended exclusively on root length. Therefore, the higher N-uptake efficiency of 'Apex' is due to its higher root-growth activity but not higher uptake rate.

A higher root growth and thus nitrate uptake activity of the N-efficient cultivar 'Apex' during the flowering and seed filling stages might be due to a better assimilate allocation to the roots (THALER and PAGES 1996). In this context the maintenance of a higher green-leaf area ('stay green') may be of particular importance (DRECCER *et al.* 2000, HORST et al. 2002). WIESLER *et al.* (2001b) illustrated higher leaf greenness (SPAD value) associated with higher photosynthetic assimilation rate for 'Apex' than for 'Capitol' during the reproductive growth stage. Although it can not yet be decided whether a higher root growth and N uptake is the consequence or the reason for 'stay green' we conclude from our results that differences in root growth particularly during the reproductive growth stage is of special importance for nitrogen efficiency in oilseed rape.

References

ANDERSON, E.L., 1987: Corn root growth and distribution as influenced by tillage and nitrogen fertilization. *Agron. J.* 79, 544-549.

BARACLOUGH, P.B., 1989: Root growth, macronutrient uptake dynamics and soil fertility requirements of a high yielding winter oilseed rape crop. *Plant Soil* 119, 59-70.

BEHRENS, T., 2002: Stickstoffeffizienz von Winterraps (Brassica napus L.) in Abhängigkeit von der Sorte sowie einer in Menge, Zeit und Form variierten Stickstoffdüngung. PhD. Thesis, University of Hannover.

BLOOM, A.J., 1997: Interaction between inorganic nitrogen nutrition and root development. *Z. Pflanzenernähr. Bodenk.* 160, 253-259.

COMFORT, S.D.; MALZER, G.L.; BUSCH, R.H., 1988: Nitrogen fertilization of spring wheat genotypes: influence on root growth and soil water depletion. *Agron. J.* 80, 114-120.

DRECCER, M.F.; SCHAPENDONK, A.H.C.M.; SLAFER, G.A.; RABBINGE, R., 2000: Comparative response of wheat and oilseed rape to nitrogen supply: absorption and utilisation efficiency of radiation and nitrogen during the reproductive growth stages determining yield. *Plant Soil* 220, 189-205.

DREW, M.C.; SAKER, L.R.; ASHLEY, T.W., 1973: Nutrient supply and the growth of the seminal root in barley: I. The effect of nitrate concentration on the growth of axes and laterals. *J. Exp. Bot.* 24, 1189-1202.

DURIEUX, R.P; KAMPRATH, E.J.; ACKSON, W.A.; MOLL, R.H., 1994: Root distribution of corn: The effect of nitrogen fertilization. *Agron. J.* 86, 958-962.

FORDE, B.L.; LORENZO, H., 2001: The nutritional control of root development. *Plant Soil* 232, 51-68.

GAO, S.; PAN, W.L.; KOENIG. R.T., 1998: Integrated root system age in relation to plant nutrient uptake activity. *Agron. J.* 90, 505-510.

GRAMI, B.; LA CROIX, L.J., 1977: Cultivar variation in total nitrogen uptake in rape. *Can. J. Plant Sci.* 57, 619-624.

GRANATO, T.C.; RAPER, C.D., 1989 Proliferation of maize roots in response to localized supply of nitrate. *J. Exp. Bot.* 40, 263-275.

HORST, W.J.; BEHRENS, T.; HEUBERGER, H.; KAMH, M.; REIDENBACH, G.; WIESLER, F., 2002: Genotypic differences in nitrogen use efficiency in crop plants. In Innovative soil-plant systems for sustainable agricultural practices. Lynch, M.J., Schepers, S. J., and Ünver, I. (Eds.). OECD. International workshop 3-7 June 2002. Izmir, Turkey.

KAPPEN, L.; SCHULTZ, G.; GRULER, T.; WIDMOSER, P., 2000: Effect of N-fertilization on shoot and roots of rape (*Brassica napus* L.) and consequence for the soil matric potential. *J. Plant Nutr. Soil Sci.* 163, 481-489.

MERRILL, S.D.; TANAKA, D.L.; HANSON, J.D., 2002: Root length growth of eight crop species in haplustoll soils. *Soil Sci. Soc. Am. J.* 66, 913-923.

OIKEH, S.O.; KLING, J.G.; HORST, W.J.; CHUDE, V.O.; CARSKY, R.J., 1999: Growth and distribution of maize roots under nitrogen fertilization in plinthite soil. *Field Crop Res.* 62, 1-13.

REIDENBACH, G.; HORST, W.J., 1997: Nitrate-uptake capacity of different root zones of *Zea mays* (L.) in vitro and in situ. *Plant Soil* 196, 295-300.

SATTELMACHER, B.; THOMS, K., 1995: Morphology and physiology of the seminal root system of young maize (Zea mays L.) plants as influenced by a locally restricted nitrate. *Z. Pflanzenernähr. Bode*nk. 158, 493-497.

STEINGROBE, B.; SCHMID, H.; CLASSEN, N., 2000: The use of the ingrowth core method for measuring root production of arable crops - influence of soil conditions inside the ingrowth core root growth. *J. Plant Nutr. Soil Sci.* 163, 617-622.

THALER, P.; PAGES, L., 1996: Root apical diameter and root elongation rate of rubber seedling (Hevea brasiliensis) show parallel responses to photoassimilate availability. *Physiol. Plant.* 97, 365-371.

THALER, P.; PAGES, L., 1998: Modelling the influence of assimilate availability on root growth and architecture. *Plant Soil* 201, 307-320.

WIESLER, F.; HORST, W.J., 1993: Differences among maize cultivars in the utilization of soil nitrate and the related losses of nitrate through leaching. *Plant Soil* 151, 193-203.

WIESLER, F.; BEHRENS, T.; HORST, W.J., 2001 a:. Nitrogen efficiency of contrasting rape ideotypes. In Horst, W. J. *et al.* (Eds.) Plant Nutrition - Food Security and Sustainability of Agro-Ecosystems. Kluwer Academic Publishers, Dordrecht Boston London. pp 60-61.

WIESLER, F.; BEHRENS, T.; HORST, W.J., 2001 b: The role of nitrogen efficient cultivars in sustainable agriculture. Optimizing Nitrogen Management in Food and Energy Production and Environmental Protection: Proceedings of the 2nd International Nitrogen Conference on Science and Policy. The Scientific World (on-line journal, www.the scientificworld.com/publications/NitrogenToC.asp.)

ZAGAL, E., 1994: Carbon distribution and nitrogen partitioning in a soil-plant system with barley (*Hordeum vulgare* L.), ryegrass (*Lolium perenne*) and rape (*Brassica napus* L.) grown in a $^{14}CO_2$-atmosphere. *Plant Soil* 166, 63-74.

Prozessregulation in der Rhizosphäre
13. Borkheider Seminar zur Ökophysiologie des Wurzelraumes
Hrsg.: W. Merbach, B. W. Hütsch, L. Wittenmayer, J. Augustin
B. G. Teubner Stuttgart, Leipzig, Wiesbaden (2003), S. 41-47

Effect of tree age on fine root biomass of *Fagus sylvatica* L. in a forest ecosystem in central Germany

Annelies CLAUS[1] and Eckhard GEORGE[2]
[1] Institute of Plant Production Sciences, Humboldt University, Invalidenstr. 42, D-10115 Berlin
[2] Institute of Vegetable and Ornamental Crops (IGZ), Theodor-Echtermeyer-Weg 1, D-14979 Großbeeren

Abstract

We investigated fine root biomass and fine root length in four different stands of a forest ecosystem in central Germany. The beech (*Fagus sylvatica* L.) trees in the different stands were 15+150, 40, 70 and 120 years old. Standing fine root biomass in these stands was 589 g m^{-2}, 684 g m^{-2}, 247 g m^{-2} and 330 g m^{-2}, respectively. Fine root biomass was almost always significantly higher in the younger forest stands (15+150 y and 40 y) than in the older stands (70 y and 120 y). In all age classes, there were relatively few roots in the organic soil layer. The largest amount of fine roots was found in the top 10 cm of the mineral soil. Specific root length was not affected by tree age.

Introduction

Fine roots are an important and dynamic component of forest biogeochemical cycling (reviews: VOGT *et al.* 1996, GILL and JACKSON 2000). Many studies on fine roots focus on climatic differences between sites (STEELE *et al.* 1997), site characteristics and nutrient availability (ABER *et al.* 1985, PERSSON *et al.* 1995). In addition to spatial variability, many studies also consider temporal variability, for example in terms of changes of fine root biomass within one growing season (ABER *et al.* 1985, HENDRICK and PREGITZER 1993). In contrast, relatively little is known about long-term effects concerning temporal variability of fine root biomass, for example in terms of changes during the life cycle of a forest stand.
The present study is part of a larger project where root biomass, turnover and belowground carbon pools are estimated in forests of different ages. The primary objective of the present study was to determine standing fine root biomass in four forest stands differing from each other in tree age only. The hypothesis was that standing biomass is lower in the younger stands (as suggested in GRIER *et al.* (1981) for an *Abies amabilis* L. stand in the Washington Cascade Mountains). A second objective was to find out whether fine root depth distribution and fine root

specific root length (SRL; root length / root biomass) are also affected by tree age. GRIER *et al.* (1981) suggested that younger forests have relatively more fine root biomass located in the top part of the mineral soil.

Material and Methods

The forest site under study, Leinefelde, is located in central Germany (lat. 51° 20' N, long. 10° 22' E, alt. 440 m above sea level). The main tree species at the Leinefelde site is *Fagus sylvatica* L., which is, at different locations and to a various extent mixed with *Acer* and *Fraxinus*. The tree age ranges from 20 to 150 years, and the investigated areas are age-homogeneous 40-, 70- and 120-year-old stands, and a 15-year-old juvenile stand which also contains several 150-year-old trees from the earlier tree generation. According to the German classification system, the soils are to be classfied as a transition between „Braunerde" and „Parabraunerde" (brown-earth), and for the 15+150-, 40-, 70-, and 120-year-old stand, they are 70-90 cm, 30-100 cm, 80 cm and 40-75 cm deep.
Fine root (< 2 mm) biomass was estimated by core sampling (PERSSON 1978) in April 2002 (n = 12). Samples were taken randomly, since the sites appeared to be homogeneous, and several studies suggested no correlation between fine root biomass and the distance to the nearest tree (JANSSENS *et al.* 1999, STOBER *et al.* 2000). A soil corer (inner diameter of 8 cm; Eijkelkamp, The Netherlands) was used to take soil samples down to a depth of 30 cm under the organic soil layer. Several studies suggested that approximately 85% of all fine root biomass is located in the top 30 cm of the soil (PERSSON *et al.* 1995, JANSSENS *et al.* 1999). The organic layer was sampled separately, followed by three intact 10 cm-increments.
The samples were stored for up to 100 days at 4°C. These storage conditions probably did not affect root biomass significantly (SMILAUEROVA 1992). Three individual samples were pooled to one sample for root washing, to reduce time-consuming laboratory work. Samples were then washed by hand, on a 500 µm sieve (BÖHM 1979), and sorted into two diameter classes (0-1 mm and 1-2 mm). The total root length of each sample was calculated after scanning the fresh roots and applying root-length measurement software (Zeiss KS 400 Version 3.0, Carl Zeiss Vision GmbH, Jena, Germany). Dry biomass was determined after drying the samples for 48 h at 70°C. Root data were calculated per unit soil surface area. Replicated measurements within a forest stand were treated as true applications in statistical analysis.

Results and Discussion

Fig. 1 shows fine root biomass from the four tree age classes of the Leinefelde forest stand. The percentage of the 1-2 mm roots of the total fine root biomass was 26% and 36% for the 70- and 120-year-old stands respectively (significantly different at the 0.05-level; Mann-Whitney U-test; MATTHYSEN 1999). In the further presentation, only the total fine root biomass is considered.

The age class had a significant influence on total fine root biomass ($F = 6.84$; $P < 0.01$). Fine root bio-mass of the younger tree stands (15+150- and 40-year-old) was almost always significantly higher than fine root biomass of the older tree stands (70- and 120-year old; Fig. 1 and Tab. 1). This is in contrast to the results published by GRIER *et al.* (1981).

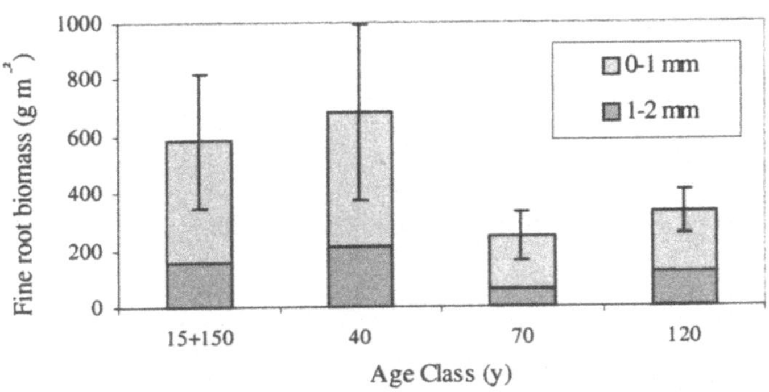

Fig. 1. Fine root (< 2 mm) biomass from the four tree age classes of the Leinefelde forest. Data are originating from one sampling occasion in April 2002. Fine roots were subdivided into the diameter classes 0-1 mm and 1-2 mm. Error bars indicate the standard deviation from the mean value of *total* fine root biomass ($n = 4$).

Tab. 1. Significance of differences between age classes (15+150-, 40-, 70- and 120-year-old stand), for the total amount of fine root biomass, normalized with a log-transformation, expressed in g/m². Mean values for the 15+150-, 40-, 70- and 120-year-old stands are 589 g m^{-2}, 684 g m^{-2}, 247 g m^{-2}and 330 g m^{-2}, respectively. P-values, calculated with the Tukey HSD test (STATISTICA for Windows, StatSoft Inc., Tulsa, OK, USA). ($n = 4$).

	Age class (y)			
	15+150	40	70	120
15+150	-	0.803	0.047	0.159
40	0.803	-	0.010	0.035
70	0.047	0.010	-	0.880
120	0.159	0.035	0.880	-

Total fine root length was 3487 m m^{-2}, 4009 m m^{-2}, 1472 m m^{-2} and 2122 m m^{-2} for the 15+150-, 40-, 70- and 120-year-old stands, respectively. Statistical differences between stands were similar as for root biomass. An explanation for the increased amount of fine roots in younger stands could be that younger trees have a higher nutrient demand.

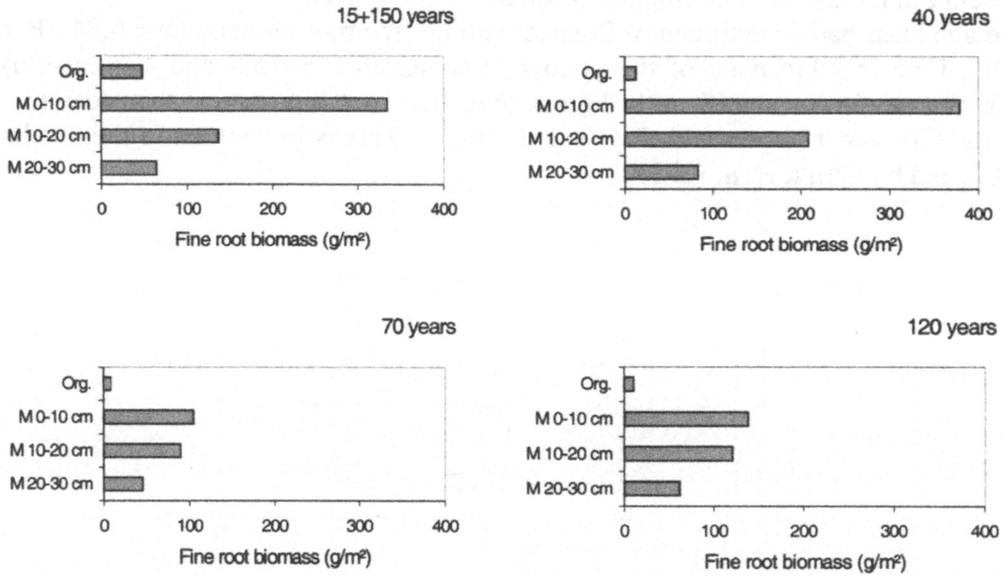

Fig. 2. Depth distribution of fine root (< 2 mm) biomass from four tree age classes of the Leinefelde forest. Data are originating from one sampling occasion in April 2002. Org., M 0-10 cm, M 10-20 cm and M 20-30 cm = organic soil layer, and 0-10, 10-20 and 20-30 cm mineral soil depth.

The depth distribution of the fine root biomass was not clearly affected by tree age (Fig. 2). There were very few roots in the thin organic layer. The majority of fine roots was located in the mineral soil layer of 0-10 cm depth and there was a gradual decrease of fine root biomass with depth in mineral soil (Fig. 3). This pattern is broadly in line with depth distributions of fine root biomass in other forests, described, for example, in GRIER *et al.* (1981) and PERSSON (1980). The relatively small amount of fine roots in the organic layer in the present study is certainly related to the fact that the organic layer in the Leinefelde forest is very thin (mean values were 1.17 cm, 0.73 cm, 1.42 cm and 1.25 cm for the 15+150-, 40-, 70- and 120-year-old stands, respectively).

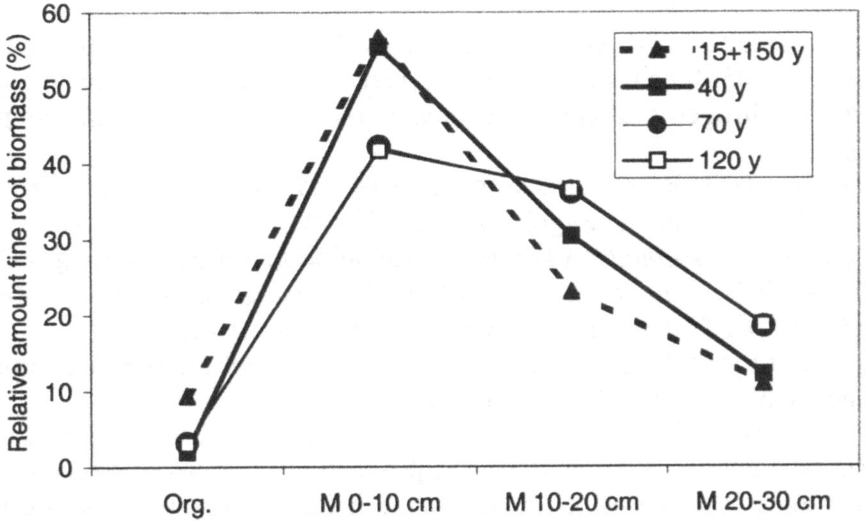

Fig. 3. Vertical distribution of relative amounts of fine root (< 2 mm) biomass (in %) from four tree age classes of the Leinefelde forest. Data are originating from one sampling occasion in April 2002. Org., M 0-10 cm, M 10-20 cm and M 20-30 cm = organic soil layer, and 0-10, 10-20 and 20-30 cm mineral soil depth.

The difference in fine root biomass between tree age classes is mainly due to differences in the top 10 cm of the mineral soil. The difference between the 15+150- and the 120-year-old stand was significant at the 0.05-level (Mann-Whitney U-test). The difference between the 15+150- and the 70-year-old stand, as well as the difference between the 40-year-old stand and both older stands, was significant at the 0.10-level only.

SRL (specific root length = root length / root biomass) was 5.91 m/g, 5.82 m/g, 5.85 m/g and 6.35 m/g for the 15+150-, 40-, 70- and 120-year-old stands, respectively. Tree roots may have a morphological pattern that is more affected by soil environmental conditions than by tree age. This was confirmed by ATKINSON (2000), who recorded similar low SRL values for different dicotyledenous species, for example 5 m/g for apple trees.

Conclusion

The present study shows that tree age has a significant influence on standing fine root biomass. Whether related processes as fine root turnover and changes in belowground carbon pools are also affected by tree age, is currently being investigated. Similar studies should be set up in various forests before general conclusions can be drawn.

References

ABER, J. D.; MELILLO, J. M.; NADELHOFFER, K. J.; MCCLAUGHERTY, C. A.; PASTOR, J., 1985: Fine root turnover in forest ecosystems in relation to quantity and form of nitrogen availability: a comparison of two methods. *Oecologia* 66, 317-321.

ATKINSON, D., 2000: Root characteristics: why and what to measure. In: A. L. Smit, A. G. Bengough, C. Engels, M. Van Noordwijk, S. Pellerin, S. C. Van de Geijn (eds.). *Root methods. A handbook.* Heidelberg: Springer-Verlag, 1-32.

BÖHM, W., 1979: Techniques of root washing. In: *Ecological Studies 33. Methods of Studying Root Systems.* Heidelberg: Springer, chapter 11.

GILL, R. A.; JACKSON, R. B., 2000: Global patterns of root turnover for terrestrial ecosystems. *New Phytologist* 147, 13-31.

GRIER, C. C.; VOGT, K. A.; KEYES, M. R.; EDMONDS, R. L., 1981: Biomass distribution and above- and below-ground production in young and mature *Abies amabilis* zone ecosystems of the Washington Cascades. *Canadian Journal of Forest Research* 11, 155-167.

HENDRICK, R. L.; PREGITZER, K. S., 1993: The dynamics of fine root length, biomass, and nitrogen content in two northern hardwood ecosystems. *Canadian Journal of Forest Research* 23, 2507-2520.

JANSSENS, I. A.; SAMPSON, D. A.; CERMAK J.; MEIRESONNE L.; RIGUZZI F.; OVERLOOP S.; CEULEMANS, R., 1999: Above- and belowground phytomass and carbon storage in a Belgian Scots pine stand. *Annals of Forest Science* 56, 81-90.

MATTYSEN, E., 1999: Biostatistiek (deel 1). *Cursusnota's 1ste licentie Biologie, Universitaire Instelling Antwerpen* (in Dutch).

PERSSON, H., 1978: Root dynamics in a young Scots pine stand in central Sweden. *Oikos* 30, 508-519.

PERSSON, H., 1980: Spatial distribution of fine-root growth, mortality and decomposition in a young Scots pine stand in central Sweden. *Oikos* 34, 77-87.

PERSSON, H.; VON FIRCKS, Y.; MAJDI, H.; NILSSON, L. O., 1995: Root distribution in a Norway spruce (*Picea abies* (L.) Karst.) stand subjected to drought and ammonium-sulphate application. *Plant and Soil* 168-169, 161-165.

SMILAUEROVA, M., 1992: Survival of short roots stored in the laboratory. In: L. Kutschera, E. Hübl, E. Lichtenegger, H. Persson, M. Sobotik (eds.). *Root ecology and its practical application. 3. ISRR Symp. Wien, Univ. Bodenkultur.* Klagenfurt, 610-612.

STEELE, S. J.; GOWER, S. T.; VOGEL, J. G.; NORMAN, J. M., 1997: Root mass, net primary production and turnover in aspen, jack pine and black spruce forests in Saskatchewan and Manitoba, Canada. *Tree Physiology* 17, 577-587.

STOBER, C.; GEORGE, E.; PERSSON, H., 2000: Root growth and response to nitrogen. In: E.-D. Schulze (ed.). *Ecological studies 142. Carbon and nitrogen cycling in European forest ecosystems.* Heidelberg: Springer-Verlag, 99-121.

VOGT, K. A.; VOGT, D. J.; BOON, P.; O'HARA, J.; ASBJORNSEN, H., 1996: Review of root dynamics in forest ecosystems grouped by climate, climatic forest type and species. *Plant and Soil* 187, 159-219.

2

Mikroökologische und molekularbiologische Wechselwirkungen in der Rhizosphäre

Prozessregulation in der Rhizosphäre
13. Borkheider Seminar zur Ökophysiologie des Wurzelraumes
Hrsg.: W. Merbach, B. W. Hütsch, L. Wittenmayer, J. Augustin
B. G. Teubner Stuttgart, Leipzig, Wiesbaden (2003), S. 51-54

Das Vorkommen von Maytansin in Celastraceen: Ist die Biosynthese des Maytansins eine Stoffwechselleistung der Pflanze oder eines wurzelbürtigen Mikroorganismus?

Eckhard LEISTNER
Institut für Pharmazeutische Biologie der Rheinischen Friedrich Wilhelms-Universität Bonn, Nussallee 6, D-53115 Bonn

Abstract

Individual plants belonging to the species *Putterlickia verrucosa* (*Celastraceae*) were collected from their natural habitat in South Africa and investigated for the presence of maytansinoids and maytansine, ansamycins of high cytotoxic activity. Maytansinoids (syn. ansamytocins) were detected in some but not all individual plants. Molecular biological analysis of the DNA from plants and a cell culture of *Putterlickia verrucosa* gave no evidence for the presence of the aminohydroxybenzoatesynthase gene which is unique to the biosynthesis of aminohydroxybenzoate, a precursor of ansamycins including maytansinoids. In contrast, observations indicate that this gene may be present in microbes of the rhizosphere of Putterlickia plants. We conclude that the roots of the Putterlickia plants may be associated with maytansinoid producing microorganisms.

Zusammenfassung

Im Rahmen der Chemotaxonomie werden Naturstoffe häufig dazu benutzt, um Verwandtschaftsbeziehungen und gemeinsame evolutionäre Ursprünge von pflanzlichen Taxa zu untersuchen. Das Vorkommen gleicher oder ähnlicher pflanzlicher Naturstoffe in verschiedenen Taxa wird als Indiz für eine Verwandtschaft zwischen diesen Taxa gewertet (HEGNAUER 1992). Es gibt jedoch auch Naturstoffe, die so weit in der Natur verbreitet sind, dass das Vorkommen identischer oder ähnlicher Naturstoffe nicht als ein Hinweis für eine Verwandtschaft zwischen zwei Taxa gewertet werden kann. Solche Naturstoffe sind die maytansinoiden Verbindungen (Fig. 1) (wie zum Beispiel Maytansin, Maytanbutin, Maytanprin, Maytansinol oder Ansamitocin), die in Höheren Pflanzen (Celastraceen, Euphorbiaceen und Rhamnaceen), Moosen (Lembophyllaceen, Thuidiaceen und Neckeraceen) sowie in Bakterien (*Actinosynnema pretiosum*) vorkommen (REIDER und ROLAND 1984, LANCINI 1986).

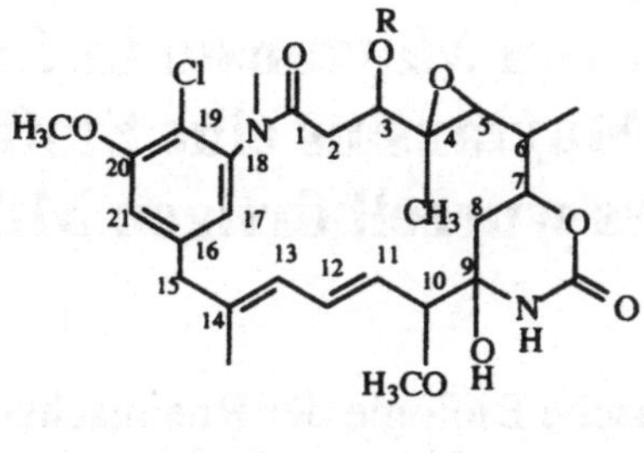

Fig. 1. Maytansinoide bakterieller (Ansamitocin P – O, P – 2, P – 3, P – 4) und pflanzlicher (Maytansinol, Maytancin, Maytansin) Herkunft: R = H Ansamitocin P – O; R = $COCH_3$ Ansamitocin P – 1; R = $COCH_2CH_3$ Ansamitocin P – 2; R = $COCH(CH_3)_2$ Ansamitocin P – 3; R = $COCH_2CH(CH_3)_2$ Ansamitocin P – 4; R = H Maytansinol; R = $COCH_3$ Maytanacin; R = $COCHCH_3NCH_3COCH_3$ Maytansin.

Für die weite Verbreitung der maytansinoiden Verbindungen gibt es zwei Erklärungsmöglichkeiten:

1. Im Zuge eines horizontalen (lateralen) Gentransfers wurden Gene der Maytansinbiosynthese intergenerisch übertragen. In der Tat gibt es über den horizontalen Gentransfer Vorstellungen (VINING 1992, BUADES und MOYA 1996), die auch auf Maytansinoide zutreffen könnten.
2. Es wäre jedoch auch denkbar, dass die in Frage kommenden Moose (s.o.) und Höheren Pflanzen von Mikroorganismen besiedelt sind, die für die Biosynthese des Maytansins verantwortlich oder mitverantwortlich sind.

In der Tat sind Höhere Pflanzen durch endophytische Pilze (CLAY 1988, CARROL 1988) und endophytische Bakterien (STURZ *et al.* 2000) besiedelt. Diese Besiedlung spielt eine erhebliche ökologische Rolle.

Bisherige Untersuchungen zur Überprüfung der unter 1 genannten Möglichkeit machten sich zunutze, dass die Gene der bakteriellen Ansamitocin (Maytansinoid)-Biosynthese seit kurzem bekannt sind (YU *et al.* 2002). Für den Fall, dass ein horizontaler (lateraler) Gentransfer während der Evolution vorgekommen ist, sollten die Gene der Ansamitocin (Maytansinoid)-Biosynthese in den Maytansinführenden Pflanzen nachweisbar sein. Um diese Frage erfolgreich zu untersuchen, wurden zwei pflanzliche Systeme, nämlich intakte Pflanzen von *Putterlickia verrucosa* (E. Meyer ex Sonder) Szyszyl sowie Zellkulturen der *Putterlickia verrucosa*-Pflanzen als Quelle für DNA ausgewählt. Oligonukleotide, die auf konservierte Regionen des Aminohydroxybenzoatsynthase-Gens gerichtet waren, wurden im Rahmen dieser Versuche eingesetzt. Das Aminohydroxybenzoatsynthase-Gen ist eines der Gene, die für die Synthese der Aminohydroxybenzoesäure zuständig sind (YU *et al.* 2002). Es kommt in den Mikroorganismen vor, die Ansa-Antibiotika bilden, also auch in *Actinosynnema pretiosum*, dem bakteriellen An-

samitocin Produzenten. Die Oligonukleotide waren an den Codon-Gebrauch der Putterlickia-Pflanzen angepasst.
Versuche mit Hilfe der Polymerase-Kettenreaktion, die unter Einsatz dieser Oligonukleotide durchgeführt wurden, ergaben in keinem Fall einen Hinweis auf das Vorkommen des Aminohydroxybenzoatsynthase-Gens in der DNA, die aus dem oberirdischen Holz oder den pflanzlichen, kultivierten Zellen isoliert wurde.
Dieser Befund ließ Zweifel an der Hypothese zum horizontalen Gentransfer aufkommen. Unterstützt wurden diese Zweifel durch eine zweite Beobachtung: Die chemische Analyse in Südafrika vorkommender, einzeln analysierter *Putterlickia verrucosa*-Pflanzen ergab, dass Maytansinoide in Extrakten einiger dieser Pflanzen nachweisbar waren, in anderen jedoch nicht (PULLEN *et al.* 2003).
Beide Befunde waren mit der Annahme erklärbar, dass Mikroorganismen für das Maytansinoid-Vorkommen Höherer Pflanzen verantwortlich sind. Die molekularbiologische Analyse der Mikroorganismen der Rhizosphäre Maytansinoidführender *Putterlicka verrucosa*-Pflanzen führten dann tatsächlich zu dem Befund, dass das Aminohydroxybenzoatsynthase-Gen in Mikroorganismen der Rhizosphäre zu finden ist (PULLEN *et al.* 2003). Diese Beobachtung leitet zu der Annahme über, dass die obengenannte Hypothese 2 richtig sein sollte und Mikroorganismen der Rhizosphäre für die Maytansinoidbiosynthese verantwortlich sind.
Wenn diese Hypothese richtig ist, dann lässt sie auch die Vermutung zu, dass Pflanzen der Celastraceen, Euphorbiaceen und Rhamnaceen, die alle der Klasse der Rosopsida angehören, Glieder einer Entwicklungslinie sind, die in der Evolution die Fähigkeit erworben und weitergegeben haben, spezifisch maytansinoidbildende Mikroorganismen in der Rhizosphäre symbiontisch zu beherbergen.

Literaturverzeichnis

BUADES, C.; MOYA, A., 1996: Phylogenetic analysis of isopenicillin-N-synthetase horizontal gene transfer. *Journal of Molecular Evolution* 42, 537-542.

CARROLL, G., 1988: Fungal endophytes in stems and leaves: from latent pathogen to mutualistic symbiont. *Ecology* 69, 2-9.

CLAY, K., 1988: Fungal endophytes of grasses: a defensive mutualism between plants and fungi. *Ecology* 69, 10-16.

HEGNAUER, R., 1992: Chemotaxonomie der Pflanzen, *Birkhäuser Verlag*, Basel.

LANCINI, G., 1986: Ansamycins. In: H. J. Rehm, G. Reed (eds), *Biotechnology*. VHC Verlagsgesellschaft, Weinheim, Vol. 4, 431-462.

PULLEN, Ch. B.; SCHMITZ, P.; HOFFMANN, D.; MEURER, K.; BOETTCHER, Th.; VON BAMBERG, D.; PEREIRA A. M.; DE CASTRO FRANCA, S.; HAUSER, M.; GEERTSEMA, H.; VAN WYK, A.; MAHMUD, T.; FLOSS, H. G.; LEISTNER, E., 2003: Occurrence and non-detectablity of maytansinoids in individual plants of the genera Maytenus and Putterlickia. *Phytochemistry* (in press).

REIDER, P. J.; ROLAND, D. M., 1984: Maytansinoids. In: *The Alkaloids*, Vol. XXIII, Academic Press Inc., 71-156.

STURZ, A. V.; CHRISTIE, B. R.; NOWAK, J., 2000: Bacterial endophytes: Potential role in developing sustainable systems of crop production. *Critical Reviews in Plant Sciences* 19, 1-30.

VINING, L. C., 1992: Secondary metabolism, inventive evolution and biochemical diversity – a review. *Gene* 115, 135-140.

YU, T.-W.; BAI, L.; CLADE, D.; HOFFMANN, D.; TOELZER, S.; TRINH, K. Q.; XU, J.; MOSS, S. J.; LEISTNER, E.; FLOSS, H. G., 2002: The biosynthetic gene cluster of the maytansinoid antitumor agent ansamitocin from *Actinosynnema pretiosum. Proceedings of the National Academy of Sciences USA* 99, 7968-7973.

Prozessregulation in der Rhizosphäre
13. Borkheider Seminar zur Ökophysiologie des Wurzelraumes
Hrsg.: W. Merbach, B. W. Hütsch, L. Wittenmayer, J. Augustin
B. G. Teubner Stuttgart, Leipzig, Wiesbaden (2003), S. 55-60

Mikrobielle Antagonisten in der Rhizosphäre: Diversität und Interaktionen

Gabriele BERG[1], Annette KRECHEL[1], Jana LOTTMANN[1], Johannes HALLMANN[2] und Kornelia SMALLA[3]
[1] Universität Rostock, Institut für Molekulare Physiologie und Biotechnologie, Mikrobiologie, Albert-Einstein-Str. 3A, D-18051 Rostock
[2] Institut für Pflanzenkrankheiten, Universität Bonn, Nußallee 9, D-53115 Bonn, Germany
[3] Biologische Bundesanstalt für Land- und Forstwirtschaft, Institut für Pflanzenvirologie, Mikrobiologie und biologische Sicherheit, Messeweg 11/12, D-38104 Braunschweig, Germany

Abstract

The rhizosphere is an important habitat for microrganisms with antagonistic properties and therefore important for plant health. Verticillium wilt caused by *Verticillium dahliae* Kleb. is a soilborne disease of many plants and responsible for high yield losses worldwide. Diversity of antagonistic bacteria associated with potential *Verticillium* host plants (potato, oilseed rape and strawberry) was studied. Regarding antibiosis against *V. dahliae*, a total of 15.642 randomly chosen bacterial isolates were screened. The percentage of isolates with antagonistic activity ranged from 3 – 47% and was influenced by the origin of isolates (plant species/developmental stage/microenviroment) and soil parameters. A total of 750 isolates were identified by their fatty acid profiles (FAME) and/or sequencing the 16S rRNA. So far, 80 antagonistic bacterial species were identified. The diversity of bacterial species was specific for each plant species and microenvironment. To assess the potential of bacterial strains in biocontrol (from this and other studies), a screening strategy using different *in vitro* and *ad planta* steps from laboratory to field trials was used. Altogether, five biological control agents were found. Based on them, three biocontrol products were developed.

Zusammenfassung

Die Rhizosphäre ist von großer Bedeutung für die Pflanzengesundheit, wobei die funktionelle Gruppe der antagonistischen Mikroorganismen eine wichtige Komponente bildet. Der bodenbürtige Pilz *Verticillium dahliae* Kleb. – der Erreger der Verticillium-Welke und gleichzeitig Modellpathogen - ist weltweit für hohe Ertragsausfälle verantwortlich. Die mikrobielle Diversität antagonistischer Bakterien von *Verticillium*-Wirtspflanzen (Kartoffel, Raps und Erdbeere) wurde in Freilandversuchen analysiert. Eine Vielzahl von Bakterien (15642) wurde überprüft. Der Anteil antagonistischer Mikroorganismen schwankte von 3–47 % und wurde durch die Herkunft der Isolate (Pflanzenspezies /Entwicklungsstadium/ Mikrohabitat) sowie von Bodenparametern beeinflusst. Nach einer Identifizierung von 750 Isolaten konnten über 80 Bakterienspezies identifiziert werden. Eine entwickelte Screeningstrategie mit verschiedenen Schritten von Labor- zu Felduntersuchungen ermöglichte eine Überprüfung der antagonistischen Bakterien hinsichtlich ihres Potenzials für die biologische Kontrolle. Hierbei wurden fünf Isolate als Biological Control Agents ausgewählt und zu 3 Produkten entwickelt.

Einleitung

Die Rhizosphäre, definiert als der an der Wurzel anhaftende und von ihr beeinflusste Boden, ist von großer Bedeutung für die Pflanzengesundheit und die Bodenfruchtbarkeit (SØRENSEN 1997). Eine wichtige Komponente bildet hierbei die funktionelle Gruppe der antagonistischen Mikroorganismen. Die Bezeichnung Antagonist <griech.> steht für Gegner oder Widersacher. Antagonistische Mikroorganismenpopulationen kommen in hohen Abundanzen in der Rhizosphäre vieler Pflanzen vor und bilden so einen natürlichen Teil der Pathogenabwehr der Pflanze gegenüber bodenbürtigen Pathogenen. In landwirtschaftlich intensiv genutzten Böden ist die mikrobielle Diversität oft stark eingeschränkt. Über eine Applikation antagonistischer Mikroorganismen kann jedoch eine wirksame biologische Kontrolle von Pathogenen erreicht werden. Diese ist von wachsender Bedeutung für eine nachhaltige Landwirtschaft. Bodenentseuchungsmittel wie Methylbromid wurden aufgrund ihrer hohen Toxizität weltweit verboten, so dass dringend Alternativen zur Bekämpfung von bodenbürtigen Pathogenen gefunden werden müssen. *Verticillium dahliae* Kleb. wird als Modellpathogen in den Untersuchungen eingesetzt. *Verticillium* ist ein bodenbürtiger Schadpilz, der weltweit hohe Ertragsausfälle an einem extrem großen Wirtspflanzenkreis verursacht. Der Pilz ist ein ideales Modellpathogen für die biologische Kontrolle bodenbürtiger Schadpilze, da er durch andere Pflanzenschutzmaßnahmen (Fungizide, Züchtung resistenter Sorten) bislang nicht bekämpfbar ist (TJAMOS 2000). In den Böden kommt es zum Aufbau eines Pathogeninokulums durch Mikrosklerotien, was langfristig zu einer exponenziellen Zunahme der Krankheit führt.

Ergebnisse und Diskussion

Mikrohabitat-spezifische Besiedlung am Beispiel der Kartoffel (*Solanum tuberosum* L.)

Die Bakterien der Kartoffel, kompartimentiert in zwei verschiedene Mikrohabitate (Phyllosphäre/Endosphäre und Rhizosphäre/Endorhiza), wurden einer komplexen Analyse mittels einer Kombination aus kultivierungsabhängigen und –unabhängigen Methoden unterzogen. Die Analyse kartoffelassoziierter Bakterien, die in zwei aufeinanderfolgenden Jahren an Freilandversuchen (Rostock und Bonn) durchgeführt wurden, zeigten eine sehr hohe Spezifität der Mikrohabitate sowie große Unterschiede zwischen endo- und ektophytisch lebenden Bakterien (KRECHEL *et al.* 2002). Die bakteriellen Gemeinschaften erwiesen sich hinsichtlich der Abundanz und Diversität als hochgradig mikrohabitatspezifisch. Bei allen untersuchten Pflanzen zeigte sich eine Veränderung innerhalb der Vegetationsperiode entsprechend des Entwicklungsstadiums der Pflanze. Die Spezifität konnte in allen bislang analysierten Proben mit kultivierungsunabhängigen Methoden bestätigt werden. Die Rhizosphäre zeichnete sich sowohl durch die höchsten mikrobiellen Abundanzen als auch durch die höchste Diversität aus. Bei den Versuchen konnte eine Vielzahl neuer *Verticillium*-Antagonisten entdeckt werden. Einige Isolate erwiesen sich als Multitarget-Antagonisten, da sie sowohl gegen phytopathogene Pilze als auch gegen den Nematoden *Meloidogyne incognita* wirksam waren. Große Unterschiede in den beiden unterirdischen Mikrohabitaten der Kartoffel wurden auch von LOTTMANN *et al.* (2000) gefunden.

Pflanzenspezifität der Rhizosphärenpopulationen

Unter Feldbedingungen wurde die Diversität der bakteriellen Gemeinschaft in der Rhizosphäre der potenziellen *Verticillium*-Wirtspflanzen Erdbeere, Raps und Kartoffel an einem Standort (Braunschweig) über zwei Jahre untersucht. Die Rhizosphäre der Pflanzen sowie der umliegende, nichtdurchwurzelte Boden wurde 3mal pro Vegetationsperiode analysiert. Mit molekularen Fingerprints konnte gezeigt werden, dass es in der Rhizosphäre zu einer pflanzenabhängigen Anreicherung bakterieller Populationen kommt und dass sich die Unterschiede in der Zusammensetzung der bakteriellen Gemeinschaft der drei Pflanzenarten im zweiten Jahr des Anbaus verstärken (SMALLA *et al.* 2001). Veränderungen in der Abundanz dominanter Bakterienpopulationen im Verlaufe der Entwicklung der Pflanzen wurden insbesondere bei jungen Pflanzen beobachtet. Obwohl viele dominante Banden in der Rhizosphäre aller Pflanzen detektiert werden konnten, gab es auch pflanzenspezifische Banden (z.B. *Nocardia* bei Erdbeere).

Eine Pflanzenspezifität konnte auch bezüglich der Abundanz und Diversität von Bakterienisolaten mit antagonistischer Wirkung gegen *Verticillium dahliae* festgestellt werden (BERG *et al.* 2002). Insgesamt wurden ca. 5000 Isolate im Dualkulturtest auf ihre antagonistische Aktivität gegenüber *Verticillium dahliae* geprüft und dabei insgesamt 286 Antagonisten isoliert. Obwohl vergleichbare Zahlen von Isolaten aus allen vier Varianten (drei Pflanzenarten und Boden) auf ihre antagonistische Aktivität hin untersucht wurden, war die Zahl der gefundenen Antagonisten für die verschiedenen Pflanzenarten sowie die Bodenprobe unterschiedlich. Die größte Zahl von Antagonisten wurde aus der Rhizosphäre der Erdbeere (125) isoliert, gefolgt von Raps (81), Kartoffel (49) und Boden (31). Alle Antagonisten wurden anhand ihrer Fettsäuremethylester-(FAME)-Muster taxonomisch charakterisiert. Die FAME-Analyse zeigte, dass ein grosser Anteil antagonistischer Isolate den Pseudomonaden in *sensu stricto* zuzuordnen war. Besonders hoch war ihr Anteil bei den von Erdbeer- und Kartoffelpflanzen isolierten Antagonisten mit knapp 90 %, während er für Raps und Boden bei 59% bzw. 55% lag. Dabei war der Anteil der als *Pseudomonas putida* B identifizierten Antagonisten aus der Erdbeer-Rhizosphäre mit 70 % besonders hoch. Die höchste Zahl zu unterschiedlichen phylogenetischen Gruppen gehörender Antagonisten („richness“) wurde aus der Rhizosphäre von Raps und Boden isoliert. Die ermittelten Diversitätsindizes „richness“ betrugen für Raps 3,3, für Boden 2,7, für Kartoffel 1,7 und für Erdbeere 1,6. Obwohl die kultiverungsunabhängige Analyse der bakteriellen Rhizosphärengemeinschaft von Raps und Kartoffel deutlich geringere Unterschiede als zur Erdbeere aufwiesen, ist das Bild bezüglich der Abundanz von Antagonisten offensichtlich anders. Typisch für Raps war die relativ große Anzahl von *Enterobacteriaceae* mit antagonistischer Aktivität gegenüber *Verticillum dahliae* (30 % der Isolate), die weder in der Rhizosphäre von Erdbeer- noch von Kartoffelpflanzen isoliert werden konnte. Da die FAME für die Identifizierung von *Enterobacteriaceae* ihre Limitierungen hat, wurde zusätzlich die 16S rDNA-Sequenzierung für Repräsentanten mit gleichen PCR-Fingerprints, die jedoch nach FAME-Analyse verschiedenen Spezies zugeordnet waren, genutzt. Von den insgesamt 46 mit FAME identifizierten Spezies wurden nur 6 aus der Rhizosphäre aller *Verticillium*-Wirtspflanzen isoliert (*P. corrugata, P. fluorescens, P. marginalis, P. putida* A/B und *Stenotrophomonas maltophilia*). Fast die Hälfte der Spezies (21 Spezies = 46 %) kam nur an einer Pflanzenart vor, allerdings wurden die Arten z.T. nur in geringer Anzahl festgestellt. Obwohl mit dem kultivierungsunabhängigen Ansatz gezeigt wurde, dass Streptomyceten zu den dominanten Bakterienpopulationen in der Rhizosphäre gehören, wurden alle Streptomyceten mit antagonistischem Potenzial gegenüber *Verticillium* aus dem Boden gewonnen.

Biologischer Pflanzenschutz mit Rhizosphärenbakterien

Eine effiziente Strategie zur Isolierung und Bewertung von bakteriellen Antagonisten als potenzielle Biological Control Agents (BCAs) wurde entwickelt (BERG *et al.* 2001b). Die traditionelle Bewertung der Antagonisten im Dualkulturtest wurde hier durch einen Klimakammertest zur Wirkung der Bakterien an Keimlingen (Raps und Erdbeere) sowie durch Untersuchungen zum molekularen Fingerprint und zum Wirkmechanismus ergänzt. In weiteren Schritten wurden die Isolate in Gewächshausversuchen und Freilandversuchen an verschiedenen, natürlich verseuchten Standorten getestet. Im Ergebnis der Untersuchungen konnten verschiedene potente BCAs herausgefunden werden, die in der Tab. 1 mit ihren Eigenschaften vorgestellt werden.

Tab. 1. Charakteristika der ausgewählten Biological Control Agents.

Spezies	Isolat	Herkunft	Potenzielle Mechanismen	Molekularer Fingerprint	Taxonomie
Bacillus subtilis[1]	B2g = DSMZ[5]8564	Raps	Antibiotika Siderophore β-1,3 Glucanase	PFGE[6] nach Verdau mit *Sfi*I	Gram-positiv
Pseudomonas fluorescens[2]	HRO-E11 = DSMZ12501	Erdbeere	Antibiotika Siderophore	BOX-PCR	Gram-negativ γ-Proteobakterien
Serratia plymuthica[2]	HRO-C48 = DSMZ12502	Raps	Chitinase (antifungische Metabolite)	PFGE[6] nach Verdau mit *Spe*I	Gram-negativ Enterics
Streptomyces albidoflavus[2]	HRO-1W1= DSMZ12500	Erdbeere	Antibiotika Siderophore Chitinase	BOX-PCR	Gram-positiv High GC
Streptomyces rimosus[3,4]	HRO-7W1 = DSMZ12424	Erdbeere	Antibiotika Siderophore Chitinase	BOX-PCR	Gram-positiv High GC

[1]BCA von Phytovit® (MARTEN *et al.* 2001)
[2]BCA von Rhizostar® (BERG *et al.* 1999)
[3]BCA von Rhizovit® (BERG und LÜTH 1999, BERG *et al.* 2001a)
[4]Identifizierung an Hand physiologischer und morphologischer Merkmale, die 16S rDNA-Sequenzierung ergab eine geringere Identität
[5]Deutsche Sammlung für Mikroorganismen und Zellkulturen Braunschweig
[6]Pulsfeldgelelektrophorese

In Zusammenarbeit mit dem Erdbeerhof Rövershagen, der *Pro*phyta Biologischer Pflanzenschutz GmbH (Malchow/Poel) und der e-nema Gesellschaft für Biotechnologie und biologischer Pflanzenschutz mbH (Raisdorf) wurden insgesamt drei Präparate zur Bekämpfung von *Verticillium dahliae* (Rhizostar) und weiteren pilzlichen Pathogenen (Rhizovit® / Phytovit®) entwickelt.

Literaturverzeichnis

BERG, G.; KURZE, S.; DAHL, R., 1999: Rhizobakterienisolate zur Anwendung gegen phytopathogene Bodenpilze und Verfahren zur Anwendung der Rhizobakterienisolate. (Isolated Rhizobacteria for Treatment of Phytopathogenic Fungal Diseases). Europäisches Patent Nr. 98124694.5.

BERG, G.; LÜTH, P., 1999: Das Rhizobakterienisolat *Streptomyces rimosus* HRO71 zur Pflanzenstärkung und Anwendung gegen phytopathogene Bodenpilze an verschiedenen Kulturpflanzen. Deutsches Patent 199-28-690.6.

BERG, G.; MARTEN, P.; MINKWITZ, A.; BRÜCKNER, S., 2001a: Efficient biological control of fungal plant diseases by *Streptomyces rimosus* DSMZ 12424. *Journal of Plant Disease and Protection* 108, 1-10.

BERG, G.; FRITZE, A.; ROSKOT, N.; SMALLA, K., 2001b: Evaluation of potential biocontrol rhizobacteria from different host plants of *Verticillium dahliae* Kleb. *Journal of Applied Microbiology* 91, 963-971.

BERG, G.; ROSKOT, N.; STEIDLE, A.; EBERL, L.; ZOCK, A.; SMALLA, K., 2002: Plant-dependent genotypic and phenotypic diversity of antagonistic rhizobacteria isolated from different *Verticillium* host plants. *Applied and Environmental Microbiology* 68, 3328-3338.

KRECHEL, A.; FAUPEL, A.; HALLMANN, J.; ULRICH, A.; BERG, G., 2002: Potato-associated bacteria and their antagonistic potential towards plant pathogenic fungi and the plant pathogenic nematode *Meloidogyne incognita* (Kofoid and White) Chitwood. *Canadian Journal of Microbiology*, in press.

LOTTMANN, J.; HEUER, H.; SMALLA, K.; BERG, G., 2000: Establishment of introduced antagonistic bacteria in the rhizosphere of transgenic potatoes and their effect on the bacterial community. *FEMS Microbiology Ecology* 33, 41-49.

MARTEN, P.; BRÜCKNER, S.; BERG, G., 2001: Biologischer Pflanzenschutz mit Rhizobakterien – eine umweltfreundliche Alternative zur Kontrolle boden- und samenbürtiger Pilze. *Gesunde Pflanzen* 53, 224-234.

SMALLA, K.; WIELAND, G.; BUCHNER, A.; ZOCK, A.; PARZY, J.; ROSKOT, N.; HEUER, H.; BERG, G., 2001: Bacterial bulk and rhizosphere communities studied by denaturing gradient gel electrophoresis of PCR-amplified fragments of 16S rRNA genes - plant-dependent enrichment and seasonal shifts revealed. *Applied and Environmental Microbiology* 67, 4742-4751.

SØRENSEN, J., 1997: The rhizosphere as a habitat for soil microorganisms. p. 21-45. In: J.D. Van Elsas, J.T. Trevors and E.M.H. Wellington (eds) *Modern soil microbiology*. Marcel Dekker, Inc. New York.

TJAMOS, E. C.; ROWE, R. C.; HEALE, J. B.; FRAVEL, D. B., 2000: Advances in *Verticillium* research and disease management. APS Press. The American Phytopathological Society St. Paul, Minnesota, 357 S.

Prozessregulation in der Rhizosphäre
13. Borkheider Seminar zur Ökophysiologie des Wurzelraumes
Hrsg.: W. Merbach, B. W. Hütsch, L. Wittenmayer, J. Augustin
B. G. Teubner Stuttgart, Leipzig, Wiesbaden (2003), S. 61-65

Microbial diversity analysis of soil by SSCP fingerprinting technique using TGGE Maxi System

Andreas LIEBER, Bärbel KIESEL und Wolfgang BABEL
Department of Environmental Microbiology, UFZ - Centre for Environmental Research Leipzig-Halle, Permoserstraße 15, D-04318 Leipzig, Germany

Abstract

Cultivation independent techniques are useful tools in analysing microbial community composition. For searching for influences of the insecticide Chlorpyrifos on bacterial communities in soil we used a modified SSCP (single strand conformation polymorphism) analysis, one of the molecular fingerprint techniques. SSCP fingerprints of Klebsiella/Serratia partial 16S rDNA sequences (V4-V5 region) were analysed in two steps both using the TGGE Maxi System (Biometra, Germany) as an example. At the first step we used the perpendicular temperature gradient function of the TGGE System for the rapid optimisation of the electrophoresis temperature. In a second step the SSCP gel was run on the TGGE Maxi System by using the optimised conditions. A good separation of distinct bands was reached. Furthermore we got a first hint on differences in community structure between treated and untreated soil samples.

Introduction

Little is known about the relationship between plants, microorganisms and animals in soil with and without anthropogenic influences. By analysing the long term influence of the insecticide Chlorpyrifos on the natural succession of a fallow field SCHÄDLER (2001) found an alteration of the dominance hierarchy of plant species caused by prevention of root herbivory. ENDLWEBER (personal communication) found a strong decrease of the population density of Collembola. The species composition of the Collembola community is altered too. Furthermore SCHÄDLER found 30% more nitrate in treated soils, but no changes in total nitrogen content of these soils. This could be an evidence for an influence on plant root exudates and lysates in soil, one of the nutrition sources for microorganisms like fungi and bacteria, or changes in microbial activity.

Since Chlorpyrifos application influences plants and insects the question arises whether or not also microorganisms are affected. For that purpose appropriate methods are necessary. The cultivation is not the first choice. Cultivation-independent analysis techniques are more useful tools for characterising microorganisms in natural habitats. Many of these techniques are based on extracted nucleic acids.

Two fingerprinting techniques are well established in community analysis – T/DGGE (temperature/denaturing gradient gel electrophoresis) and SSCP (single strand conformation polymorphism analysis). We preferred the SSCP for analysing fragments amplified from 16S rRNA genes in complex bacterial communities. For this purpose, PCR is conducted with a non-phosphorylated and a 5′-phosphorylated primer. Both primers bind to conserved regions in the small subunit (SSU) rRNA genes. Single-stranded DNA is obtained by lambda exonuclease digestion of the phosphorylated strands of the double stranded PCR products. By this means, heteroduplex formation during electrophoresis, which cannot be avoided in T/DGGE methods, is eliminated (SCHWIEGER and TEBBE 1998). On the other hand, the optimisation of electrophoresis conditions is more comfortable with TGGE. We performed SSCP on the TGGE System to use advantages of both methods. The occurrence of Klebsiella/Serratia in the field was analysed as a model.

Methods

Soil samples (30 g, 2-7 cm depth) were collected from a long term field experiment nearby Bad Lauchstädt (middle Germany) in early August. Total soil DNA was extracted directly by a bead beating procedure (modified protocol, FastDNA® SPIN Kit, BIO-101, USA), CTAB (cethyltrimethylammonium-bromide) extraction and PEG (polyethylene glycol) precipitation. Soil samples (10 g each plot, 10 identical plots per treating manner) were pooled and a slurry was made by shaking it with 100 ml ice cold aqua bidest. at 200 rpm for 5 minutes. Quintuplicate aliquots of 200 mg of each soil suspension, according to 100 mg soil, were added to Lysing Matrix E tubes supplied with the FastDNA® SPIN Kit for soil. 978 μl sodium phosphate buffer (kit supplied), 122 μl MT-buffer (kit supplied) and 10 μl • -mercaptoethanol per tube were added. The samples were processed for 30 seconds at speed 5.5 m/sec. in the FastPrep® Instrument (BIO-101, USA). Lysing matrix E tubes were centrifuged at 14,000 x g for 5 minutes. The supernatants were extracted twice with CTAB until no white interface is visible (MURRAY and THOMPSON 1980). The DNA was precipitated with 2 volumes 30% PEG-6000 (Fluka BioChemika)-1.6 M NaCl (GRIFFITHS *et al.* 2000) within two hours at room temperature. Samples were centrifuged at 18,000 x g for 10 minutes to pellet the DNA. The DNA pellet was washed twice with 70% ethanol, air dried and resolved in 100 μl ultra pure water. The E.Z.N.A. Cycle Pure Kit (Peqlab, Germany) was used for final purification. The DNA concentration and ratio A_{260}/A_{230} and A_{260}/A_{280} were determined spectrophotometrically. The DNA preparations were stored at –20°C.

For amplification of variable regions (V4-V5) of 16S rRNA genes primers (Tab. 1) were evaluated and the Klebsiella/Serratia specific primer was designed using the ARB package (http://www.arb.de.vu/), the Ribosomal database project (http://www.rdp.cme.msu.edu/html/), the BLASTN program (http//www. ncbi.nlm.nih.gov/BLAST/) and the evaluation version of the Primer Premier 5 program (http://www.premierbiosoft.com/).
PCR was done with PCR Taq polymerase Master Mix Kit (Qiagen, Germany) and additional Mg^{2+}. The first PCR was performed with a group specific and a "universal" 16S rDNA primer within 30 cycles. The final concentration of each primer was 0.5 μM. We used 1-10 ng template DNA in 20 μl reaction volume. A second semi-nested PCR contained two "universal" 16S rDNA primers and 0.06 μl product of the first PCR reaction in a 60 μl reaction volume. There the reverse primer was 5´-phosphorylated. This second PCR was performed within 17 cycles. The final concentration of each primer was 0.5 μM. PCR amplifications were conducted in a thermocycler using the following conditions: initial denaturation for 3 min at 94°C followed by 30 or 17 cycles of denaturing at 94°C for 20 sec, annealing at 53 or 58°C (see Tab. 1) for 30 sec, extension at 72°C for 60 sec, and a final primer extension for 7 min completes this PCR. The semi-nested PCR product was purified with the E.Z.N.A. Cycle Pure Kit (Peqlab, Germany). The DNA was eluted in 30 μl ultra pure water.
To obtain single strand DNA for SSCP analysis the purified PCR product was treated with 12-U-λ-exonuclease (NEB, USA) at 37°C for 60 min. The samples containing single-stranded DNA were mixed with one volume of SSCP loading buffer (95% formamide, 10 mM NaOH, 0.025% bromophenol blue), heated to 95°C for 2 min and then immediately chilled on ice for at least 5 min. These samples can be stored at –20°C for several months.
SSCP analysis was performed on the TGGE Maxi System (Biometra, Germany) under the following conditions: gel matrix of 0.6 x MDE stock solution (Biozym, Germany), 1 x TBE-buffer (Sigma, Germany), 400 Volt, maximum current, 16-20 hours, and a constant temperature between 20-30°C. For the optimisation of the electrophoresis temperature we used the perpendicular temperature gradient function of the TGGE System. In a previous run a temperature gradient ranging from 15 to 30°C was applied perpendicular to the current direction.
After the run gels were stained by a silver staining procedure described by BASSAM et al. (1991) with the modification that the developing solution consist of 3% NaOH and 0.2% formaldehyde [37% wt/vol]. The stained gel was then incubated in conservation solution (10% ethanol, 13% glycerol) for 30 min and dried between the support film and a cellophane sheet at 60°C for 4 hours. The gels can be stored in the dark for several months.

The dried gels were transferred into 24-bit bitmap image files (flatbed scanner Umax PowerLook 2000) and converted into 8-bit grey scale images using Corel Photo Paint 8.

Tab. 1. Primers that have been used for microbial community analysis.

Primer name	Primer sequence (5´-3´)	Targeted SSU rRNA gene	T_a (°C)	References
Kleb-Serr94f	GTG ACG AGC GGC GGA C	Klebsiella/ Serratia	58	designed, this work
Uni-Bac927r	CCC GTC AAT TYM TTT GAG TT	bacteria		mod.SCHWIEGER and TEBBE 1998
Uni-Bac515f	GTG CCA GCA GCC GC	bacteria	53	mod.SCHWIEGER and TEBBE 1998
Uni-Bac927r	CCC GTC AAT TYM TTT GAG TT	bacteria		mod.SCHWIEGER and TEBBE 1998

Results

From control soil and soil after Chlorpyrifos application (a chernozem - silt loam, humus-rich, dark colour)in the total DNA extracts yields of approximately 30 μg DNA per g (dry weight) were obtained. The size of extracted DNA ranged from 8 to 20 kbp. The DNA extracts were colourless and subsequent PCR reactions were not inhibited. Typical absorbance ratios were 1.36 (A_{260}/A_{230}) and 1.7 (A_{260}/A_{280}). SSCP fingerprints of Klebsiella/Serratia partial 16S rDNA sequences (V4-V5 region) were analysed in two steps. At the first step the optimal temperature for SSCP fingerprints was determined as 26°C (Fig. 1). At this temperature around 43 individual, well separated bands were visible, among them 18 sharp bands and a few faint bands. At higher or lower temperature fewer bands are distinguishable. Repeated extractions of individual 100 mg aliquots of the pooled samples resulted in similar fingerprints. In a second step the SSCP was run with ssDNA PCR products from the control areas and after application of Chlorpyrifos in five parallels (Fig. 2). The two different profiles contained the same intense bands, but the relative intensities of some bands (e.g. Fig. 2, band A) varied between the samples. The intensity of one band (Fig. 2, band B) was decreased in all samples of Chlorpyrifos treated soil. This could be a first hint for shifts in the bacterial community under the influence of Chlorpyrifos treatment. And this will enable us to widen the model for studying the whole community for a deeper insight into changes in the bacterial community.

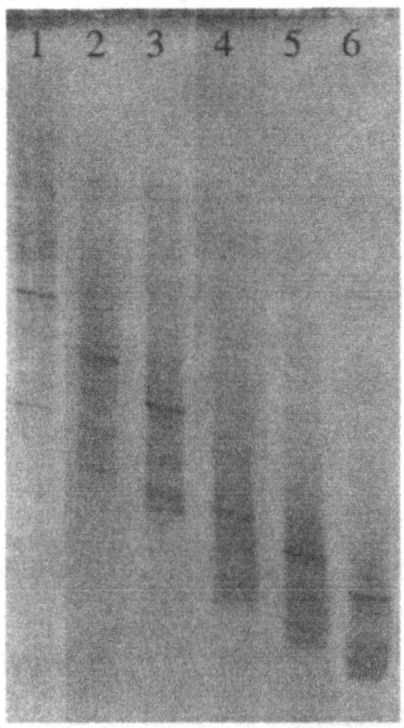

Fig. 1. Perpendicular temperature gradient SSCP of a control sample (K3) ; lane 1 15°C; lane 2 17°C; lane 3 20°C, lane 4 23°C; lane 5 26°C; lane 6 29°C.

band A

band B

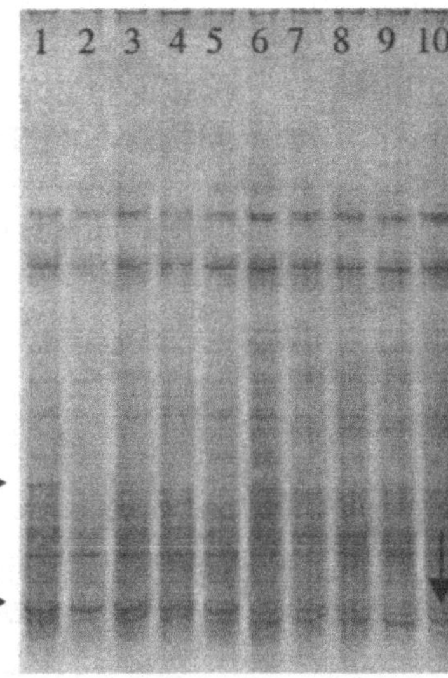

Fig. 2. SSCP of samples of untreated (lanes 1-5) and treated soil (lanes 6-10); the band B indicates a band that is decreased in all treated soil samples.

References

BASSAM, B. J.; CAETANO-ANOLLES, G.; GREEHOFF, P. M., 1991: Fast and sensitive silver staining of DNA in polyacrylamide gels. [erratum in Analytical Biochemistry 198, 217] to *Analytical Biochemistry* 196, 80-83.

GRIFFITHS, R. I.; WHITELEY, A. S.; O'DONNELL, A. G.; BAILEY, M. J., 2000: Rapid method for coextraction of DNA and RNA from natural environments for analysis of ribosomal DNA and rRNA-based microbial community composition. *Applied and Environmental Microbiology* 66, 5488-5491.

MURRAY, M. G.; THOMPSON, W. F., 1980: Rapid isolation of high-molecular-weight plant DNA. *Nucleic Acids Research* 8, 4321-4325.

SCHÄDLER, M., 2001: Einfluß phytophager Insekten auf die Struktur und Dynamik einer Ackerbrache (Dissertation). – *UFZ - Berichte* 11/2001, 1 – 111.

SCHWIEGER, F.; TEBBE, C. C., 1998: A new approach to utilize PCR–single-strand-conformation polymorphism for 16S rRNA gene-based microbial community analysis. *Applied and Environmental Microbiology* 64, 4870-4876

Prozessregulation in der Rhizosphäre
13. Borkheider Seminar zur Ökophysiologie des Wurzelraumes
Hrsg.: W. Merbach, B. W. Hütsch, L. Wittenmayer, J. Augustin
B. G. Teubner Stuttgart, Leipzig, Wiesbaden (2003), S. 66-67

Mikrobielle Ökologie von „autoinducer" - produzierenden Bakterien in der Rhizosphäre von Tomatenpflanzen

Stephan GANTNER[1], Christine DÜRR[1], Michael SCHMID[1], Frank B. DAZZO[2] und Anton HARTMANN[1]
[1]Institut für Bodenökologie, Abteilung Rhizosphärenbiologie, Forschungszentrum für Umwelt und Gesundheit, Ingolstädter Landstr.1, D-85764 Neuherberg
[2]Department of Microbiology, Michigan State University, East Lansing, USA

Abstract

Gram-negative bacteria have got a special kind of cell-cell-communication controlled by free-diffusible pheromone-like compounds called N-Acyl-L-Homoserine lactones (AHL). We investigated the effect of AHL-producing bacteria in the rhizosphere of MicroTom tomato plants, in axenic and soil systems using wild type strains of *Serratia liquefaciens* MG1 and *Pseudomonas putida* Isolate F as well as their AHL-negative mutant strains.

Zusammenfassung

Gram-negative Bakterien verfügen über ein eigenes Regulationssystem, das sie zur Interaktion befähigt (EBERL 1999). Diese Art der Zell-Zell-Kommunikation erfolgt über Signalmoleküle vom Typ N-Acyl-L-Homoserin-Lacton (AHL), welche erst ab Erreichen eines bestimmten Schwellenwertes eine zellspezifische Reaktion auslösen (*Quorum sensing*), z.B. die Produktion von Antibiotika oder Enzymen (SWIFT *et al.* 1996). AHL-produzierende Rhizosphärenbakterien sind damit u.a. im Hinblick auf mögliche Abwehrmechanismen gegen Phytopathogene bei Pflanzen von besonderem Interesse.

Wir untersuchten das Verhalten der AHL-produzierenden Bakterienstämme *Serratia liquefaciens* MG1 und *Pseudomonas putida* IsoF, sowie deren AHL-Negativmutanten (AHL^-), die kein AHL mehr produzieren konnten, an der Rhizosphäre von Tomate. Beide Stämme zeigen eine AHL-regulierte Eigenschaft zur Resistenz gegen verschiedene Pilze. Zur besseren *in* situ –Lokalisierung wurden die verwendeten Stämme *gfp*- und *rfp*- markiert eingesetzt. Als Versuchsmodell diente eine kleinwüchsige Tomatenpflanze in Blumentopferde als Bodensystem bzw. in PhytaTray®-Boxen mit Kristallsand als steriles (axenisches) System (SIMONS *et al.* 1996). Mittels FISH (Fluoreszenz *in situ*-Hybridisierung) und CLSM (Confokale Laser Scanning Mikroskopie) Studien konnte gezeigt werden,

dass die AHL-Produktion keinen Einfluß auf die Wachstumsdichte oder -art der untersuchten Bakterien hat. Die Zellzahl der inokulierten Bakterien an den Wurzeln stabilisierte sich nach einer Woche auf 10^{6-7} Zellen/g Frischgewicht. Das Besiedlungsverhalten zeigte sich ebenfalls unbeeinflusst von der AHL-Produktion im Vergleich der Wildtypen (Produzenten) und der AHL-Negativmutanten.
Zur Überprüfung der AHL-Produktion *in situ* wurden Monitorstämme konstruiert, die selbst AHL-Negativmutanten waren, jedoch auf die Anwesenheit von AHL mit spezifischen Reaktionen antworteten (z.B. mit rfp /gfp Produktion) (STEIDLE *et al.* 2001). Es konnte mit Hilfe des CMEIAS (Center for Microbial Ecology-Image-Analysis-System) in Verbindung mit der CLSM eine AHL-Produktion der Wildtypen an der Wurzel nachgewiesen werden.
Mittels des t-RFLP (terminaler Restriktions-Längen-Polymorphismus) wurde ferner eine Veränderung der bakteriellen Rhizosphärenpopulation nach Inokulation von *Serratia liquefaciens* (Wildtyp) beobachtet, während die inokulierten Bakterienstämme selbst nach einwöchiger Inkubation keine Dominanz in der Zusammensetzung der Rhizosphärenpopulation zeigten.

Besonderer Dank gilt den beteiligten Projektpartnern:
PD Dr. Christian Langebartels, Regina Schuhegger, der Arbeitsgruppe der pflanzlichen Toxikologie, des Biochemischen Pflanzenpathogenitätinstuts der GSF und PD Dr. Leo Eberl, sowie Anette Steidle der Arbeitsgruppe „quorum-sensing“ der Technischen Universität München.

Literaturverzeichnis

EBERL, L., 1999: N-Acyl homoserinelactone-mediated gene regulation in gram-negative bacteria. System. *Applied Microbiology* 22, 493-506.

SIMONS, M.; VAN DER BIJ, A. J.; BRAND, I.; DE WEGER, L. A.; WIJFFELMAN, C. A.; LUGTENBERG, J. B. J., 1996: Genotobiotic system for studying rhizosphere colonization by plant growth-promoting *Pseudomonas* bacteria. *Molecular Plant-Microbe Interactions* 9, 600-607.

STEIDLE, A.; SIGL, K.; SCHUHEGGER, R.; GANTNER, S.; IHRING, A.; SCHMID, M.; STOFFELS, M.; RIEDEL, K.; GIVSKOV, M.; HARTMANN, A.; LANGEBARTELS, C.; EBERL, L., 2001: Visualization of *N*-Acylhomoserine Lactone (AHL)-mediated cell-cell communication between bacteria colonizing the tomato rhizosphere. *Applied and Environmental Microbiology* 67, 5761-5770.

SWIFT, S.; THROUP, J. P.; WILLIAMS, P.; SALMOND, G. P. C.; STEWART, G. S. A. B., 1996: Quorum sensing a population-density component in the determination of bacterial phenotype. *Trends in Biochemical Science* 21, 214-219.

Prozessregulation in der Rhizosphäre
13. Borkheider Seminar zur Ökophysiologie des Wurzelraumes
Hrsg.: W. Merbach, B. W. Hütsch, L. Wittenmayer, J. Augustin
B. G. Teubner Stuttgart, Leipzig, Wiesbaden (2003), S. 68-74

Einfluss von Bodenbakterien auf Transkriptom und Proteom des Mykorrhizapilzes *Amanita muscaria*

Mika T. TARKKA, Andreas MAIER und Rüdiger HAMPP
Eberhard-Karls-Universität Tübingen, Botanisches Institut, Physiologische Ökologie der Pflanzen, Auf der Morgenstelle 1, D-72076 Tübingen

Abstract

Microscopy of ectomycorrhizal roots directly sampled from the field has revealed the presence of bacteria in the hyphal sheath, in the Hartig net region and on soil-colonising extramatrical hyphae. Some of these rhizosphere bacteria, named mycorrhiza helper bacteria, have been shown to improve hyphal growth and mycorrhiza formation. The mechanisms behind these processes are currently unknown. We have focused on the effect of two streptomycete strains, that induce hyphal growth and mycorrhization, on the gene and protein expression of *Amanita muscaria* (fly agaric). Total RNA isolated from *A. muscaria* mycelium, co-cultured with the streptomycete strain 505, has been used for cDNA synthesis and for the construction of a subtracted cDNA bank. Differential expression of the subtracted cDNAs has been analysed by macroarrays and partially verified by Northern hybridisation. Changes in protein patterns between streptomycete-induced *A. muscaria* mycelium and of aseptic fungal cultures have been analysed by two-dimensional gel electrophoresis and peptide mass spectrometry. Most of the differentially expressed fungal genes show homology to members of signal transduction pathways. Others are homologous to genes involved in metabolism, cell proliferation and cell structure. This implies that the fungus-streptomycete interaction is connected with changes in signalling pathways that lead to an altered fungal physiology.

1. Mykorrhiza-Helfer-Bakterien

Die meisten Landpflanzen bilden Symbiosen mit filamentösen Pilzen. Bäume und Sträucher in unseren Breitengraden gehen überwiegend die sogenannte Ektomykorrhiza (ECM) ein. Diese ist gekennzeichnet durch die Entwicklung eines Pilzgeflechts um die Feinwurzeln der Pflanzen sowie durch Ausbildung eines Pilzmyzels zwischen Epidermis und Rindenzellen (Hartigsches Netz). Der Stoffaustausch zwischen den Symbiosepartnern findet im Bereich des Hartigschen Netzes statt. Dabei versorgt die Pflanze den Pilz mit Kohlenhydraten und erhält im Gegenzug vom Pilz Wasser, Nährsalze und Aminosäuren (SMITH und READ 1997). Die Entwicklung und die Funktion der Ektomykorrhiza ist bis heute nur ansatzweise auf molekularer Ebene untersucht (MARTIN 2001, VOIBLET *et al.* 2001). Während der Entwicklung der Ektomykorrhiza-Symbiose kommt es zu morphologischen Veränderungen der Pilzhyphen. Dies äußert sich in Form angeschwollener und stark verzweigter Hyphen. Es gilt als wahrscheinlich, dass die Interaktionen zwischen Pilz und Pflanze zu Modifikationen in der Organisation des Zytoskeletts führen (RAUDASKOSKI *et al.* 2001). Als mögliche Regulatoren dieses Entwicklungsprozesses wurden bisher zwei Gene identifiziert: Ein bei der Symbiose hochreguliertes *Ras*-Gen (SUNDARAM *et al.* 2001) und ein differentiell lokalisiertes Cdc42-Homolog (GORFER *et al.* 2001).

Neben den Interaktionen zwischen Pilz und Pflanze wird die Mykorrhiza-Bildung maßgeblich von verschiedenen Bodenbakterien beeinflusst. Mikroskopische Untersuchungen zeigten Bakterien auf dem Pilzmantel, im Hartigschen Netz, sowie auf dem Myzelium im Boden (FOSTER und MARKS 1966, NURMIAHO-LASSILA *et al.* 1997). Bei der Bildung der Mykorrhiza werden Stoffe in den Boden ausgeschieden, was zu einer Änderung der Mikrobenpopulationen führt (KAZNELSON *et al.* 1962, TIMONEN *et al.* 1998). Die Mikroorganismen in dieser sogenannten „Mykorrhizosphäre" erfüllen diverse Aufgaben und können verschiedene Eigenschaften der Mykorrhizapilze komplementieren. Einige Bakterienstämme sind zur Stickstoff-Fixierung befähigt (SECILIA und BAGYARAJ 1987), andere an der Mineralisierung von Phosphat beteiligt (TORO *et al.* 1996). Manche Stämme können das Wachstum von Pflanzen anregen (CHANWAY 1997, POOLE *et al.* 2001, PROBANZA *et al.* 2001) oder das Wachstum pathogener Organismen hemmen (BUDI *et al.* 1999, SBRANA *et al.* 2000, WHIPPS 2001).

Neben ihrer Wirkung auf Pflanzen beeinflussen die Bakterien auch das Pilzwachstum und die Mykorrhizierung. BOWEN und THEODOROU (1979) haben als erste gezeigt, dass einige Bakterienarten *in vitro* das Hyphenwachstum und die Bildung von Ektomykorrhizen induzieren. Diese Bakterien wurden später als fluoreszierende *Pseudomonaden* und sporulierende *Bacilli* identifiziert (GARBAYE und BOWEN 1989, SCHELKLE und PETERSON 1996). Schon 1994 hatte GARBAYE (1994) für diese Bakterien den Namen „Mycorrhiza Helper Bacteria" (MHB) vorgeschlagen. Kürzlich wurden auch *Burkholderia, Paenibacillus* und *Rhodococcus*-Arten als MHB identifiziert (POOLE *et al.* 2001). Einige MHB reagieren nur mit spezifischen Pilzarten, so dass davon ausgegangen wird, dass sich spezifische Erkennungssysteme zwischen MHB und Pilzen entwickelt haben (GARBAYE und DUPONNOIS 1992). Der molekulare Hintergrund dieser Wechselwirkungen ist jedoch völlig unklar.

2. Streptomyceten-Stämme 505 und 1003

Um MHB von *Amanita muscaria* zu charakterisieren, haben wir Bakterien aus Arealen, in denen unsere Modellpilze vorkommen, isoliert. Mehrere der Bakterienisolate fördern sowohl das Wachstum von *Amanita muscaria* als auch die Mykorrhizierung. Im Gegensatz dazu hemmen sie die Entwicklung von pathogenen Pilzen. Für molekulare Analysen wurden zwei Streptomycetenstämme ausgewählt, der Stamm 1003, der bei *A. muscaria* am stärksten wachstumsinduzierend wirkt, und der Stamm 505, der schwächer wachstumsfördernd wirkt. Die Morphologie der *A. muscaria*-Hyphen ändert sich deutlich durch die Interaktion. Die induzierten Hyphen sind dünner und wachsen weiter von einander entfernt als die nicht induzierten (MAIER, PORALLA UND HAMPP, unveröffentlicht).

Streptomyceten stellen einen wichtigen Teil der Bakterienflora im Waldboden dar, meist werden sie wegen ihrer Antibiotika-Biosynthese untersucht. CRAWFORD *et al.* (1993) charakterisierten sie jedoch auch als potentielle Antagonisten von pathogenen Pilzen. Hierbei kann die Unterdrückung des Pilzwachstums durch Streptomyceten oft ebenfalls auf die Produktion von Sekundärmetaboliten zurückgeführt werden (HWANG *et al.* 2001).

3. Veränderte Protein- und Genexpressionsmuster von *A. muscaria* bei Co-Kultur mit Streptomyceten

Im Proteinmuster auf zweidimensionalen Gelen von *A. muscaria*-Hyphen treten in Anwesenheit der Bakterien Veränderungen auf. Durch Massenspektrometrie konnte ein in seiner Expression verringertes Protein als Aktin identifiziert werden (MAIER, unveröffentlicht). Darüber hinaus wurde eine subtraktive cDNA-Bibliothek aus morphologisch veränderten *A. muscaria*-Hyphen, die über einen Zeitraum von zehn Wochen zusammen mit dem Stamm 505 kultiviert worden waren, erstellt (TARKKA, unveröffentlicht). Hiermit sollen Komponenten möglicher Signalketten kloniert und so der proteinchemische Ansatz auf Ebene der Transkription komplementiert werden. Komponenten der Signalketten sind durch proteinchemische Techniken aufgrund ihrer niedrigen Expressionsrate nur schwer zu identifizieren.

In diesem ersten Ansatz konnte eine Vielzahl von differentiell exprimierten Pilzgenen identifiziert werden. Hierzu wurden die verschiedenen cDNAs aus dem subtraktiven Ansatz auf Nylonmembranen transferiert und mit Gesamt-cDNA aus Pilzkulturen, die mit bzw. ohne Bakterienstamm 505 kultiviert wurden, hybridisiert. Fünfzig von 220 untersuchten subtrahierten cDNAs zeigen einen deutlichen Unterschied der Genexpression bei der Streptomyceten-Pilz-Interaktion. Die meisten Gene wurden durch die Streptomyceten-Stämme 505 und 1003 induziert. Die Resultate wurden teilweise durch Northern-Hybridisierung bestätigt.

Durch partielle Sequenzierung und Vergleich der abgeleiteten Proteinsequenzen konnten für 28 cDNAs, die deutliche Unterschiede im Genexpressionsmuster zeigten, Homologien mit bekannten Proteinen gefunden werden. Dabei zeigte sich, dass viele in ihrer Expression durch Streptomyceten modulierten Gene mit Komponenten homolog sind, die an Signalketten (Src-ähnliche Protein-Kinase, Tyrosin-Protein-Kinase) mitwirken bzw. zur Regulation der Genexpression (Tab. 1) dienen. Weiterhin waren Enzyme (wie z.B. Mono-Oxidase), Zellzyklus-Proteine (Histon) sowie Proteine der Zellwandbildung (z.B. Agglutinin) dabei. Diese Resultate deuten darauf hin, dass in Co-Kultur eine Veränderung in der Signaltransduktion von *A. muscaria* stattfindet, welche die Physiologie des Pilzes deutlich verändert. Für spezifische Genexpressionsanalysen wurden die 3'-UTR von vorerst drei cDNA-Fragmenten, welche bei *A. muscaria - Populus tremula/tremuloides* cDNA Bank-Analysen (NEHLS und PÜHLER, unveröffentlicht) nicht gefunden werden konnten, durch RACE isoliert.

Tab. 1. Beispiele für Komponenten verschiedener Signalkettenmodule, deren Expression bei Co-Kultur von *A. muscaria* mit Streptomyceten verändert wird.

Gene mit veränderter Expression in Co-Kultur	Mögliche Funktion
Ser/Thr kinase (MAPKKK)	Protein phosphorylation
Tyrosine protein kinase	Protein phosphorylation
Mss4p	Phosphatidylinositol-4-phosphate 5-kinase
Rab-interacting protein	Regulation of secretion
Transcription factor	Regulation of transcription
Transcription factor	Activation of transcription

Molekulare Untersuchungen zu Bakterien-Pilz-Interaktion wurden bisher lediglich an dem Ascomyceten *Aspergillus nidulans* durchgeführt. Hier hemmt die Zugabe des Antibiotikums Bafilomycin von *Streptomyces halstedii* (K122) das Wachstum der Pilzhyphen und führt zu zusätzlichen Verzweigungen. Von den bei dieser Interaktion in ihrer Expression veränderten Genen sind bislang zwei charakterisiert worden, eines, das für ein induzierbares Zellwand-Protein kodiert, und ein in seiner Expression reprimiertes Gen, das für das ASPDN1 Glykoprotein kodiert (MELIN *et al.* 1999).

Unsere Arbeiten zeigen deutlich, dass die pilzliche Genexpression während der Interaktion mit Streptomyceten stark verändert ist. Viele der bisher isolierten, differentiell exprimierten Gene und Proteine sind vermutlich für die Veränderung der Physiologie der Pilze im Zuge der Interaktion mit den Bakterien von Bedeutung. Ein weiteres Ziel des Projektes ist die Charakterisierung von Genen in *Amanita muscaria,* die in ihrer Expression in frühen Stadien der Interaktion verändert werden. Dadurch sollen mögliche Signalkettenkomponenten identifiziert werden, die auf Bakterien-Exsudate reagieren.

Literaturverzeichnis

BOWEN, G. D.; THEODOROU, C., 1979: Interactions between bacteria and ectomycorrhizal fungi. *Soil Biology and Biochemistry* 11, 119-126.

BUDI, S. W.; VAN TUINEN, D.; MARTINOTTI, G.; GIANINAZZI, S., 1999: Isolation from the *Sorghum bicolor* mycorrhizosphere of a bacterium compatible with arbuscular mycorrhiza development and antagonistic towards soilborne fungal pathogens. *Applied and Environmental Microbiology* 65, 5148-5150.

CRAWFORD, D. L.; LYNCH, J. M.; WHIPPS, J. M.; OUSLEY, M. A., 1993: Isolation and characterization of actinomycete antagonists of a root fungal pathogen. *Applied and Environmental Microbiology* 59, 3899-3905.

CHANWAY, C. P., 1997: Inoculation of tree roots with plant growth promoting soil bacteria: an emerging technology for reforestation. *Forest Science* 43, 99-112.

FOSTER, R. C.; MARKS, G. C., 1966: The fine structure of the mycorrhizas of *Pinus radiata*. *Australian Journal of Biological Science* 19, 1027-1038.

GARBAYE, J., 1994: Helper bacteria: a new dimension to the mycorrhizal symbiosis. *New Phytologist* 128, 197-210.

GARBAYE, J.; BOWEN, G. D., 1989: Stimulation of mycorrhizal infection of *Pinus radiata* by some microorganisms associated with the mantle of ectomycorrhizas. *New Phytologist* 112, 383-388.

GARBAYE, J.; DUPONNOIS, R., 1992: Specificity and function of mycorrhization helper bacteria (MHB) associated with the *Pseudotsuga menzii-Laccaria laccata* symbiosis. *Symbiosis* 14, 335-344.

GORFER, M.; TARKKA, M. T.; HANIF, M.; PARDO, A.; LAITIAINEN, E.; RAUDASKOSKI, M., 2001: Characterization of small GTPases Cdc42 and Rac and the relationship between Cdc42 and actin cytoskeleton in vegetative and ectomycorrhizal hyphae of *Suillus bovinus*. *Molecular Plant Microbe Interactions* 14, 135-144.

HWANG, B. K.; LIM, S. W.; KIM, B. S.; LEE, J. Y.; MOON, S. S., 2001: Isolation and in vivo and in vitro antifungal activity of phenylacetic acid and sodium phenylacetate from *Streptomyces humidus*. *Applied and Environmental Microbiology* 67, 3739-3745.

KAZNELSON, H.; ROUATT, J. W.; PETERSON, E. A., 1962: The rhizosphere effect of mycorrhizal and non-mycorrhizal roots of yellow birch seedlings. *Canadian Journal of Botany* 40, 377-382.

MELIN, P.; SCHNURER, J.; WAGNER, E. G. H., 1999: Changes in *Aspergillus nidulans* gene expression induced by bafilomycin, a *Streptomyces*-produced antibiotic. *Microbiology* 145, 1115-1122.

MARTIN, F., 2001: Frontiers in molecular mycorrhizal research - genes, loci, dots and spins. *New Phytologist* 150, 499-507.

NURMIAHO-LASSILA, E. L.; TIMONEN, S.; HAAHTELA, K.; SEN, R., 1997: Bacterial colonization patterns of intact *Pinus sylvestris* mycorrhizospheres in dry pine forest soil: An electron microscopy study. *Canadian Journal of Microbiology* 43, 1017-1035.

PROBANZA, A.; MATEOS, J. L.; LUCAS GARCIA, J. A.; RAMOS, B.; DE FELIPE, M. R.; GUTIERREZ MANERO, F. J., 2001: Effects of inoculation with PGPR *Bacillus* and *Pisolithus tinctorius* on *Pinus pinea* L. growth, bacterial rhizosphere colonization, and mycorrhizal infection. *Microbial Ecology* 41, 140-148.

POOLE, E. J.; BENDING, G. D.; WHIPPS, J. M.; READ, D. J., 2001: Bacteria associated with *Pinus sylvestris-Lactarius rufus* ectomycorrhizas and their effects on mycorrhiza formation in vitro. *New Phytologist* 151, 743-751.

RAUDASKOSKI, M.; PARDO, A.; TARKKA, M.; GORFER, M.; HANIF, M.; LAITIAINEN, E., 2001: Small GTPases, cytoskeleton and signal transduction in filamentous homobasidiomycetes. In: A. Geitmann, M. Cresti, I Brent Heath (eds), *Cell biology of plant and fungal tip growth.* NATO Science Series I, 328, 123-136.

SBRANA, C.; BAGNOLI, G.; BEDINI, S.; FILIPPI, C.; GIOVANNETTI, M.; NUTI, M. P., 2000: Adhesion to hyphal matrix and antifungal activity of *Pseudomonas* strains isolated from *Tuber borchii* ascocarps. *Canadian Journal of Microbiology* 46, 259-268.

SCHELKLE, M.; PETERSON, L., 1996: Suppression of common root pathogens by helper bacteria and ectomycorrhizal fungi *in vitro*. *Mycorrhiza* 6, 481-485.

SECILIA, J.; BAGYARAJ, D. J., 1987: Bacteria and actinomycetes associated with pot cultures of vesicular-arbuscular mycorrhizas. *Canadian Journal of Microbiology* 33, 1069-1073.

SMITH, S. E.; READ, D. J., 1997: *Mycorrhizal symbiosis*. Academic Press, London, England.

SUNDARAM, S.; KIM, S. J.; SUZUKI, H.; MCQUATTIE, C. J.; HIREMAH, S. T.; PODILA, G. K., 2001: Isolation and characterization of a symbiosis-regulated ras from the ectomycorrhizal fungus *Laccaria bicolor*. *Molecular Plant Microbe Interactions* 14, 618-628.

TIMONEN, S.; JORGENSEN, K. S.; HAAHTELA, K.; SEN, R., 1998: Bacterial community structure at defined locations of *Pinus sylvestris-Suillus bovinus* and *Pinus sylvestris-Paxillus involutus* mycorrhizospheres in dry pine forest humus and nursery peat. *Canadian Journal of Microbiology* 44, 499-513.

TORO, M.; NEDIALKOVA, K.; AZCON, R.; BAREA, J. M., 1996: Establishment of two rock phosphate solubilizing bacteria in the rhizosphere of mycorrhizal onion plants and their effect on plant growth in a microcosm, pp. 665-668. In: C. Azcon-Aguilar, and J. M. Barea (ed.), *Mycorrhizas in integrated systems: from genes to plant development* (EUR 16728). European Commission, Luxembourg.

VOIBLET, C.; DUPLESSIS, S.; ENCELOT, N.; MARTIN, F., 2001: Identification of symbiosis regulated genes in *Eucalyptus globulus-Pisolithus tinctorius* ectomycorrhiza by differential hybridization of arrayed cDNAs. *Plant Journal* 25, 181-191.

WHIPPS, J. M., 2001: Microbial interactions and biocontrol in the rhizosphere. *Journal of Experimental Botany* 52, 487-511.

Prozessregulation in der Rhizosphäre
13. Borkheider Seminar zur Ökophysiologie des Wurzelraumes
Hrsg.: W. Merbach, B. W. Hütsch, L. Wittenmayer, J. Augustin
B. G. Teubner Stuttgart, Leipzig, Wiesbaden (2003), S. 75-80

Physiologische und molekularbiologische Ansätze zur Analyse der Funktion von Indol-3-buttersäure in der Entwicklung arbuskulärer Mykorrhizen

Michael KALDORF[1], Dorothee FITZE[2], Raul HERRERA[3] und Jutta LUDWIG-MÜLLER[2]
[1]Institut für Ökologie der Friedrich-Schiller Universität Jena, Germany
[2]Institut für Botanik der Technischen Universität Dresden, Germany
[3]Institute of Plant Biotechnology, Talca University, Talca, Chile

Abstract

Inoculation of maize with the AM fungus *Glomus intraradices* resulted in a distinct root phenotype with a significant increase in the percentage of lateral roots during early stages of colonization. Auxins are known to play a role in different developmental processes including fine root formation and are therefore suitable candidates for the regulation of AM symbiosis. The increase in lateral root development coincided with an increase in the levels of the auxin indole-3-butyric acid (IBA). The phenotype of mycorrhizal maize roots could be mimicked by exogenously applied IBA to non-mycorrhizal roots. Addition of trifluoro-IBA (TFIBA), an inhibitor of IBA-induced root growth and lateral root induction, simultaneously with IBA resulted in a phenotype resembling that of untreated controls. In roots treated with TFIBA the inoculation with AM-fungi did not increase the formation of fine roots. The TFIBA-treatment also reduced endogenous free IBA and the percentage of colonization in mycorrhizal roots. Using activity screening and cDNA-AFLP analysis, we have started to isolate the gene for IBA synthetase and other genes differentially induced by IBA and AM.

Abkürzungen: ABA, Abscisinsäure; AM, arbuskuläre Mykorrhiza; IAA, Indol-3-essigsäure; IBA, Indol-3-buttersäure; TFIBA, 4,4,4-Trifluoro-3-(indol-3-)buttersäure

Einleitung

Arbuskuläre Mykorrhiza (AM) ist die vermutlich älteste und am weitesten verbreitete Form von Symbiose zwischen Pilzen und höheren Pflanzen. Fossile Belege deuten darauf hin, dass schon die ersten Landpflanzen Symbiosen mit AM-Pilzen eingingen (REDECKER *et al.* 2000); unter den rezenten Pflanzen bilden nach HARLEY und HARLEY (1987) ca. 70% der Arten AM-Symbiosen aus. Mykorrhizierte Pflanzen zeigen häufig ein gesteigertes Wachstum in Folge einer verbesserten Versorgung mit Nährstoffen sowie einer erhöhten Stressresistenz, während der Kohlenhydratbedarf der obligat biotrophen AM-Pilze durch die Pflanzen gedeckt wird (SMITH und READ 1997). Während der Entwicklung einer AM-Symbiose werden einerseits Arbuskeln und Vesikel der AM-Pilze innerhalb der Wurzeln gebildet; andererseits induzieren die Pilze morphogenetische Modifikationen im Wurzelsystem. Als Basis für diese Effekte werden Änderungen in der Pflanzenernährung sowie im Phytohormonhaushalt diskutiert (BERTA *et al.* 1993). Unter anderem wurden erhöhte Gehalte an Cytokinin (ALLEN *et al.* 1980) und ABA in mykorrhizierten Wurzeln gefunden, während der IAA-Gehalt unverändert war (DANNEBERG *et al.* 1992). In neueren Untersuchungen wurden Auxinstandards eingesetzt, die mit stabilen Isotopen markiert waren; zusammen mit Gaschromatographie/Massenspektroskopie als Analyseverfahren waren dadurch genaue Messungen der Gehalte verschiedener natürlicher Auxine möglich. Damit konnte gezeigt werden, daß der Gehalt eines weiteren natürlich vorkommenden Auxins, der Indol-3-buttersäure (IBA), in jungen mykorrhizierten Maiswurzeln erhöht ist (LUDWIG-MÜLLER *et al.* 1997, KALDORF und LUDWIG-MÜLLER 2000). Im folgenden werden Versuche dargestellt, bei denen durch den Einsatz des IBA-Antagonisten TFIBA der Zusammenhang zwischen IBA, Mykorrhizierung und morphogenetischen Effekten untersucht wurde. Außerdem werden molekularbiologische Ansätze vorgestellt, mit denen die Rolle der IBA in der AM-Symbiose näher untersucht werden soll.

Charakterisierung der TFIBA als IBA-Antagonist

Exogene Applikation von IBA führte bei jungen Maispflanzen zu einem signifikanten Anstieg des Sproß- und Wurzelfrischgewichtes auf das ca. Dreifache der Werte für Kontrollpflanzen. Im für das Wachstum optimalen Konzentrationsbereich von 1-10 μM IBA bildeten IBA-behandelte Pflanzen zudem deutlich mehr Lateralwurzeln als Kontrollpflanzen, während eine IBA-Konzentration von 100 μM das Wurzelwachstum hemmte (KALDORF und LUDWIG-MÜLLER 2000). Die Zugabe von TFIBA hatte bei unbehandelten Pflanzen über den Konzentrationsbereich von 1–1000 μM nur minimale negative Effekte auf das Wurzelwachstum. Dagegen wurden die positiven Effekte exogen applizierter IBA durch 10 μM TFIBA teilweise und durch 100 μM TFIBA vollständig aufgehoben. TFIBA konnte in Aufnahmeexperimenten bereits nach 10 Minuten in Wurzeln nachgewiesen werden (Daten nicht gezeigt); die antagonistische Wirkung der TFIBA muß daher nicht unbedingt auf einer reduzierten Aufnahme der zugegebenen IBA beruhen, sondern könnte auch durch ein Eingreifen von TFIBA in die Signaltransduktion erklärt werden. Darüber hinaus stellt die IBA-Synthetase, die die Synthese von IBA aus IAA und Acetyl-CoA katalysiert, wahrscheinlich einen weiteren Angriffspunkt der TFIBA dar. Die spezifische Aktivität der IBA-Synthetase war in TFIBA-behandelten Pflanzen ca. 10 mal geringer als in Kontrollen, während die Aktivität der IBA-Synthetase in zellfreien Extrakten durch TFIBA nicht beeinflußt wurde.

IBA in mykorrhizierten Maiswurzeln
Die Inokulation von Maiskeimlingen mit dem AM-Pilz *Glomus intraradices* Sy167 (freundlicherweise zur Verfügung gestellt von S. Reinhard, Institut für Pflanzenernährung, Universität Hohenheim) führte bei 10 – 20 Tage alten Maispflanzen – wie auch die Zugabe von IBA – zu einem Wurzelphänotyp, der durch einen erhöhten Anteil an Feinwurzeln charakterisiert war (Tab. 1). Gleichzeitig war in dieser Entwicklungsphase die Konzentration freier IBA in mykorrhizierten Wurzeln erhöht (Abb. 1). Beide Effekte traten nicht auf, wenn die inokulierten Pflanzen mit TFIBA behandelt wurden.

Tab. 1. Einfluß der Inokulation mit *G. intraradices* sowie der Behandlung mit 100 µM TFIBA auf die Feinwurzelentwicklung junger Maispflanzen. Angegeben ist der Anteil der Feinwurzeln in Prozent des gesamten Wurzelfrischgewichtes 11 Tage nach Inokulation. Die Daten sind Mittelwerte von je sechs Pflanzen pro Experiment ± Standardabweichung.

Exp. Nr.	Kontrolle	*Glomus*	Verhältnis *Glomus*/Kontrolle (- TFIBA)	Kontrolle + TFIBA	*Glomus* + TFIBA	Verhältnis *Glomus*/Kontrolle (+ TFIBA)
1	16 ± 5	25 ± 8	1,6	-	-	-
2	28 ± 6	42 ± 3	1,5	27 ± 13	16 ± 4	0,9
3	19 ± 4	31 ± 6	1,7	16 ± 4	18 ± 5	1,1

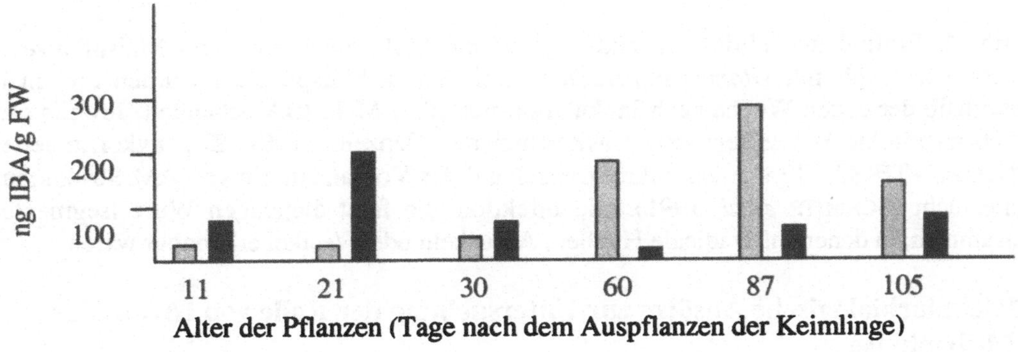

Abb. 1. Vergleich der Gehalte an freier IBA in Wurzeln mykorrhizierter und nicht mykorrhizierter Maispflanzen. Kontrollpflanzen ■ ; mit *Glomus intraradices* inokulierte Pflanzen ▨ .

Bei den mit *G. intraradices* inokulierten Pflanzen führte die Zugabe von TFIBA nicht nur zur oben beschriebenen Reversion der AM-typischen Stimulierung der Feinwurzelbildung, sondern auch zu einer weniger erfolgreichen Mykorrhizierung (Abb. 2). Der signifikant erhöhte Anteil von Pilzhyphen außerhalb der Wurzeln bei den mit TFIBA behandelten Pflanzen spricht gegen eine allgemein toxische Wirkung von TFIBA auf *G. intraradices*; der IBA-Antagonist TFIBA scheint die erfolgreiche Kolonisierung der Wurzeln durch die Mykorrhizapilze spezifisch zu hemmen. Diese Beobachtung passt zu früheren Befunden, wonach die Behandlung inokulierter Pflanzen mit IBA nicht nur zu einem gesteigerten Sproß- und Wurzelwachstum, sondern auch zu einer höheren Mykorrhizierungsrate führt (DUTRA *et al.* 1996, GUNZE und HENNESSY 1980); im Gegensatz dazu wird die Mykorrhizierung schon durch Zugabe von 1μM IAA signifikant gehemmt (GRYNDLER *et al.* 1998).

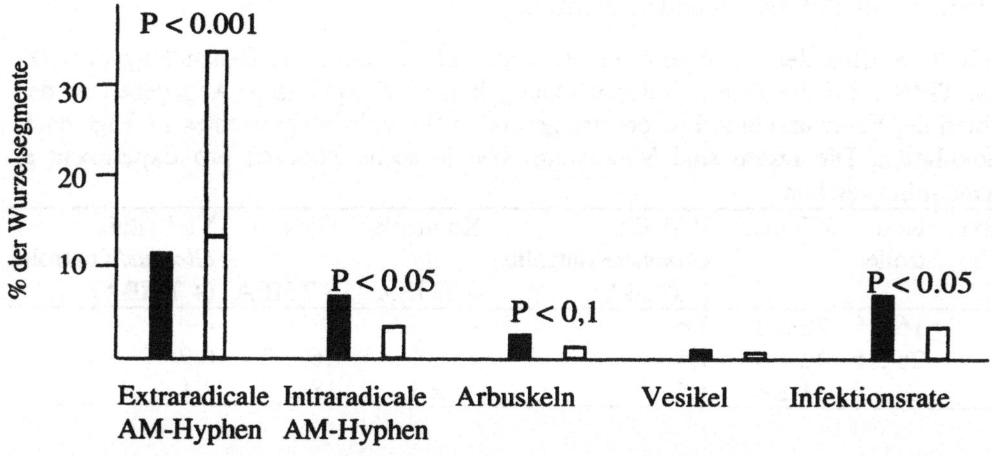

Abb. 2. Einfluß der TFIBA-Behandlung auf die Mykorrhizierung von Maispflanzen. Sechs von zwölf mit *Glomus intraradices* inokulierten Maispflanzen wurden drei mal innerhalb der ersten Woche nach Inokulation mit 100 μM TFIBA behandelt. Jeweils ca. 2500 angefärbte Wurzelsegmente (mykorrhizierte Pflanzen -TFIBA ■ ; mykorrhizierte Pflanzen +TFIBA □) wurden mikroskopisch auf das Vorhandensein von AM-Strukturen untersucht (SCHMITZ *et al.* 1991). Die Infektionsrate fasst diejenigen Wurzelsegmente zusammen, in denen intraradicale Hyphen, Arbuskeln oder Vesikel erkennbar waren.

Molekularbiologische Ansätze zur Untersuchung der Rolle von IBA in der AM-Symbiose

Einen wesentlichen Beitrag zum Verständnis der Rolle von IBA bei der Entwicklung der Mykorrhiza könnten Untersuchungen zur Expression und Lokalisation der IBA-Synthetase liefern, da dieses Enzym wahrscheinlich ein Schlüsselenzym der IBA-Synthese ist (LUDWIG-MÜLLER *et al.* 1995). Dazu sollte zunächst das Gen für

IBA-Synthetase durch Aktivitätsscreening isoliert werden. Der erste Schritt bestand dabei im Anlegen einer cDNA-Expressionsbibliothek aus vier Tage alten Maiskeimlingen. Nach heterologer Expression der cDNAs in *E. coli* wurden dann die rekombinanten Proteine aus den Bakterien isoliert und in einem einfachen Enzymassay (LUDWIG-MÜLLER *et al.* 1995) auf IBA-Synthetase-Aktivität getestet. Nachdem in einem Aliquot der cDNA-Bank eine geringe IBA-Synthetase-Aktivität nachweisbar war, wurde begonnen, einzelne Pools der Expressionsbibliothek zu untersuchen. Es sollte möglich sein, durch die Untersuchung von Einzelkolonien aus dem derzeit vorliegenden Pool mit der höchsten IBA-Synthetase-Aktivität einen einzelnen Klon zu isolieren, in dem das IBA-Synthetase-Gen aus Mais kloniert und expremiert ist.
In einem weiteren Ansatz sollten mit Hilfe von cDNA-AFLPs Maisgene identifiziert werden, deren Expression durch Kolonisierung der Wurzeln mit AM-Pilzen und durch IBA-Zugabe reguliert wird. Dazu wurden vier Wochen alte Maispflanzen verwendet, die entweder drei Wochen vor der Ernte mit *Glomus intraradices* inokuliert oder 4 bzw. 24 Stunden vor der Ernte mit 10 μM IBA behandelt worden waren. Aus diesem Material wurde RNA isoliert und zur cDNA-Synthese eingesetzt. Erste AFLP-Analysen zeigten, daß verschiedene Gene sowohl in den mykorrhizierten, als auch in den IBA-behandelten Pflanzen induziert waren. Mit der Klonierung und Identifizierung dieser Gene wurde begonnen.

Literaturverzeichnis

ALLEN, M. F.; MOORE, T. S. J.; CHRISTENSEN, M., 1980: Phytohormone changes in *Bouteloua gracilis* infected by vesicular-arbuscular mycorrhizae: I. Cytokinin increases in the host plant. *Canadian Journal of Botany* 58, 371-374.

BERTA, G.; FUSCONI, A.; TROTTA, A., 1993: VA mycorrhizal infection and the morphology and function of root systems. *Environmental and Experimental Botany* 33, 159-173.

DANNEBERG, G.; LATUS, C.; ZIMMER, W.; HUNDESHAGEN, B.; SCHNEIDER-POETSCH, H.; BOTHE, H., 1992: Influence of vesicular-arbuscular mycorrhiza on phytohormone balances in maize (*Zea mays* L.). *Journal of Plant Physiology* 141, 33-39.

DUTRA, P. V.; ABAD, M.; ALMELA, V.; AUGUSTÍ, M., 1996: Auxin interaction with the vesicular-arbuscular mycorrhizal fungus *Glomus intraradices* Schenck & Smith improves vegetative growth of two citrus rootstocks. *Scientia Horticulturae* 66, 77-83.

GRYNDLER, M.; HRSELOVA, H.; CHVATALOVA, I.; JANSA, J., 1998: The effect of selected plant hormones on *in vitro* proliferation of hyphae of *Glomus fistulo-*

sum. Biologia Plantarum 41, 255-263.

GUNZE, C. M. B.; HENNESSY, C. M. R., 1980: Effect of host-applied auxin on development of ectomycorrhiza in cowpeas. *Transactions of the British Mycological Society* 74, 247-251.

HARLEY, J. L.; HARLEY, E. L., 1987: A check-list of mycorrhiza in the British flora. *The New Phytologist* 105 (Suppl.), 1-102.

KALDORF, M.; LUDWIG-MÜLLER, J., 2000: AM fungi might affect the root morphology of maize by increasing indole-3-butyric acid biosynthesis. *Physiologia Plantarum* 109, 58-67.

LUDWIG-MÜLLER, J.; HILGENBERG, W.; EPSTEIN, E., 1995: The *in vitro* biosynthesis of indole-3-butyric acid in maize. *Phytochemistry* 40, 61-68.

LUDWIG-MÜLLER, J.; KALDORF, M.; SUTTER, E. G.; EPSTEIN, E., 1997: Indole-3-butyric acid (IBA) is enhanced in young maize (*Zea mays* L.) roots colonized with the arbuscular mycorrhizal fungus *Glomus intraradices*. *Plant Science* 125, 153-162.

REDECKER, D.; KODNER, R.; GRAHAM, L. E., 2000: Glomalean fungi from the Ordovician. *Science* 289, 1920-1921.

SCHMITZ, O.; DANNEBERG, G.; HUNDESHAGEN, B.; KLINGNER, A.; BOTHE, H., 1991: Quantification of vesicular-arbuscular mycorrhiza by biochemical parameters. *Journal of Plant Physiology* 139, 106-114.

SMITH, S. E.; READ, D. J., 1997: *Mycorrhizal symbiosis*. 2nd ed., Academic Press, San Diego, London, New York, Boston, Sydney, Tokyo, Toronto.

3

Stofftransformation und -akkumulation im Wurzelraum

Prozessregulation in der Rhizosphäre
13. Borkheider Seminar zur Ökophysiologie des Wurzelraumes
Hrsg.: W. Merbach, B. W. Hütsch, L. Wittenmayer, J. Augustin
B. G. Teubner Stuttgart, Leipzig, Wiesbaden (2003), S. 83-89

Induzierte Phytosanierung schwermetallbelasteter Böden mit Hilfe von Sonnenblumen

Anne BOLHÖFER [1], Ilka MEINE[2], Kay DOMEYER [1], Gerhard WELP [2] und Heinrich W. SCHERER [1]
[1] Institut für Pflanzenernährung der Rheinischen Friedrich-Wilhelms-Universität Bonn, Karlrobert-Kreiten-Straße 13, D-53115 Bonn; E-Mail: aci@uni-bonn.de
[2] Institut für Bodenkunde der Rheinischen Friedrich-Wilhelms-Universität Bonn, Nussallee 13, D-53115 Bonn

Abstract

Nowadays the development of alternative clean-up techniques for heavy metal contaminated soils becomes more and more important. An ecological and economical way to remove heavy metals could be via phytoremediation with heavy metal accumulating plants. Previous research focussed on a decrease of the total contents of heavy metals in soil, even though the significance of total contents is very limited from an (eco-)toxicological point of view.
In a pot experiment with sunflowers we analysed the mobility and binding forms of heavy metals over time after adding various mobilizing agents to polluted soils in order to enhance the soil-plant transfer of the metals.

Einleitung

Die Entwicklung alternativer Bodensanierungsverfahren ist sinnvoll, da kontaminierte Böden aufgrund ihrer ökologischen und ökonomischen Bedeutung zunehmend ein Problem darstellen (LOMBI *et al.* 1999). Im Vergleich zu konventionellen Sanierungsverfahren ist die Phytosanierung kostengünstiger, da die Installations- und Erhaltungskosten vergleichsweise niedrig sind. Darüber hinaus ist sie ökosystemschonend und hat auch deshalb eine hohe Akzeptanz in der Öffentlichkeit. Allerdings ist die angestrebte Sanierungszeit von fünf Jahren zur Zeit auch unter optimierten Bedingungen noch nicht erreicht (HUND und WIESNER 1999). Forschungsbedarf ist also vorhanden.
Bei der induzierten Phytosanierung wird der Transfer der Schadstoffe vom Boden in die Pflanze durch eine Zugabe mobilisierender Agenzien gefördert. In verschiedenen amerikanischen Arbeiten ist diese Technik insbesondere bei Blei wirkungsvoll eingesetzt worden (z. B. WU *et al.* 1999). Eine Schwachstelle bisheriger Sanierungsverfahren ist dabei, dass ihr Erfolg lediglich an einer Abnahme der Schwermetallgesamtgehalte gemessen wird, obwohl die Aussagekraft von Ge-

samtgehalten aus (öko-)toxikologischer Sicht sehr begrenzt ist (ALEF 1994). Ein wesentliches Ziel unserer Arbeit war deshalb, im Rahmen einer induzierten Phytosanierung die Veränderung der Schwermetall-Mobilität und -Bindungsformen in belasteten Böden über die Zeit zu verfolgen.

Material und Methoden

Versuchsböden

In Tabelle 1 sind die Kennwerte der drei Versuchsböden zusammengefasst. Ihre Schwermetallgehalte liegen deutlich über den in der Bodenschutzverordnung für die jeweilige Bodenart festgelegten Vorsorgewerten (BBODSCHV 1999). Während die Böden 1 und 2 durch anthropogene Einflüsse belastet sind, zeichnet sich Boden 3 durch eine geogene Belastung aus.

Tab. 1. Kennwerte der Versuchsböden (SM-Gehalte in mg kg^{-1}, Königswasseraufschluss).

Parameter	Boden 1	Boden 2	Boden 3
pH ($CaCl_2$)	5,2	5,8	6,3
Corg (%)	3,05	2,14	4,46
Bodenart	uL	lU	uL
Cd	2,5	16,1	12
Cu	98	58	104
Pb	326	82	877
Zn	695	436	1240

Versuchsdesign

Der Versuch wurde in Kick-Brauckmann-Gefäßen (8,8 kg Boden je Gefäß) durchgeführt. Beim Befüllen der Versuchsgefäße wurde zunächst der Gefäßboden mit 10 cm Boden bedeckt, auf den in vertikaler Anordnung zwei Diffusionstaschen (Abb. 1) gestellt wurden, bevor die Gefäße mit dem restlichen Boden befüllt wurden. Bei den Diffusionstaschen handelt es sich um PVC-Ringe mit einem Durchmesser von 9 cm und einer Tiefe von 3 cm, die nach dem Befüllen mit dem jeweiligen Versuchsboden (ca. 200 g je Tasche) beidseitig mit einer Gaze von 1 μm Maschenweite bespannt wurden. Die Versuchsböden und auch der Inhalt der Diffusionstaschen waren durch Düngung mit N, P, K und S optimal mit Nährstoffen versorgt.

Nach dem Bepflanzen wurden die Böden auf 60 % der jeweiligen maximalen Wasserkapazität eingestellt. Der Wassergehalt der Böden wurde während der Vegetationsperiode je nach Witterung und Wasserverbrauch täglich ein- bis zweimal kontrolliert und fehlendes Wasser ergänzt.

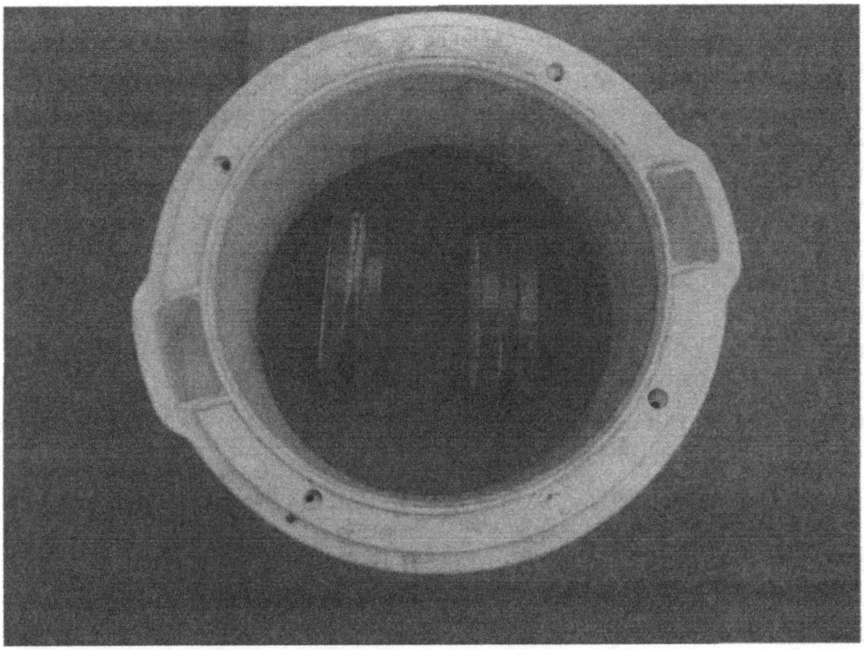

Abb. 1. Kick-Brauckmann-Gefäß mit „Taschen".

Versuchspflanze

Als Versuchspflanze diente die Sonnenblume (*Helianthus annuus*, cv. Capella). Um einheitliches Pflanzenmaterial zu erhalten, wurden die Sonnenblumen in Quarzsand vorkultiviert und zehn Tage nach dem Auflaufen in die Versuchsgefäße (5 Pflanzen je Gefäß) umgepflanzt. Durch diese Art der Anzucht sollte gewährleistet werden, dass die Pflanzen ihren Wurzelapparat in einem nicht mit Schwermetallen belasteten Boden ausbilden.

Versuchsvarianten

- Kontrolle (ohne Zugabe mobilisierender Agenzien)
- EDTA 1 (1 mmol kg^{-1} Boden) (*EDTA 1*)
- EDTA 2,5 (2,5 mmol kg^{-1} Boden) (*EDTA 2,5*)
- Citronensäure (10 mmol kg^{-1} Boden) (*Citrsre*)
- Citronensäure+EDTA (10 mmol kg^{-1} Boden+1 mmol kg^{-1} Boden)(*Citr+EDTA*)

Die Zugabe von EDTA und Citronensäure erfolgte eine Woche nach dem Umpflanzen der Sonnenblumen. Hierzu wurden die entsprechenden Mengen der oben genannten Lösungen mit 500 ml entsalztem Wasser verdünnt und anschließend gleichmäßig zur Hälfte von oben und zur Hälfte von unten zugegeben. Ein anschließendes Überbrausen der Böden sollte eine gleichmäßige Verteilung der

Agenzien gewährleisten.
Drei und sechs Wochen nach der Behandlung sowie zum Zeitpunkt der Blüte wurden die Sonnenblumen an der Bodenoberfläche abgeschnitten und in Spross und Blätter und beim letzten Erntetermin zusätzlich noch in Blüte zerlegt. Danach wurden der Boden und die Diffusionstaschen vorsichtig aus den Gefäßen entnommen. Die Gewinnung von „Rhizoboden" erfolgte durch vorsichtiges Abschütteln des in unmittelbarer Wurzelnähe anhaftenden Bodens. Diese Art der Gewinnung von Rhizoboden wird zwar in der Literatur häufig beschrieben (RILEY und BARBER 1969, HENDRIKS und JUNGK 1981), lässt aber keine hohe räumliche Auflösung zu. Der Boden aus den Diffusionstaschen wurde unmittelbar in flüssigem Stickstoff tiefgefroren, um Wasserbewegungen und damit verbundene Verlagerungen von Schwermetallen zu verhindern. Durch Abdrehen des so gefrorenen Bodenblocks mittels einer Drehbank in Schichten von 1 mm konnte schließlich Boden in definierten Abständen von der Wurzeloberfläche gewonnen werden.

Extraktion bzw. Aufschluss der Schwermetalle

- Wasserextraktion (2 g Boden, 20 ml dest. Wasser, 2 h schütteln,
- Cocktail-Extraktion (2 g Boden, 50 ml Lösung (0,025 M EDTA, 1 M NH_4OAc, 0,1 M $NH_2OH{\bullet}HCl$), 2 h schütteln, 15 min zentrifugieren bei 2500 *g*)
- Druckaufschluss der Pflanzenproben unter Verwendung von 65 %iger HNO_3.

Schwermetallmessung

Die Messung der Schwermetalle erfolgte, je nach vorliegender Konzentration, mit ICP-OES (Cd, Cu, Pb, Zn) oder GF-AAS (Cd, Pb).

Erste Ergebnisse

Die Trockenmasseerträge der Sonnenblumen bei Boden 1 sind in Abb. 2 dargestellt. Dabei fällt auf, dass die Kontrollvariante des ersten Erntetermins die höchste Trockenmasse aufweist. Die übrigen Varianten zeigen aufgrund der Behandlung, insbesondere *„EDTA 2,5"*, eine Wachstumsdepression. Zum dritten Erntetermin fällt *„EDTA 2,5"* gegenüber den anderen Varianten deutlich im Ertrag zurück, so dass diese Dosis offensichtlich phytotoxisch war. Die Behandlungen *„EDTA 1"* und *„Citrsre"* ergeben zu Versuchsende Erträge leicht oberhalb der Kontrolle, während der Ertrag bei kombinierter Gabe von EDTA und Citronensäure leicht abfällt.
Bei einem Vergleich der wasserlöslichen Cd-Gehalte zwischen Rhizoboden und Nicht-Rhizoboden (siehe Abb. 3) zeigt sich bei den Varianten *„Citrsre"* und *„Citr+EDTA"* eine deutliche Verarmung in der Rhizosphäre. Dies könnte auf die

versauernde und komplexierende Wirkung von Citronensäure zurückgeführt werden. Die Zugabe von EDTA weist unabhängig von der Konzentration nur geringe Effekte auf.

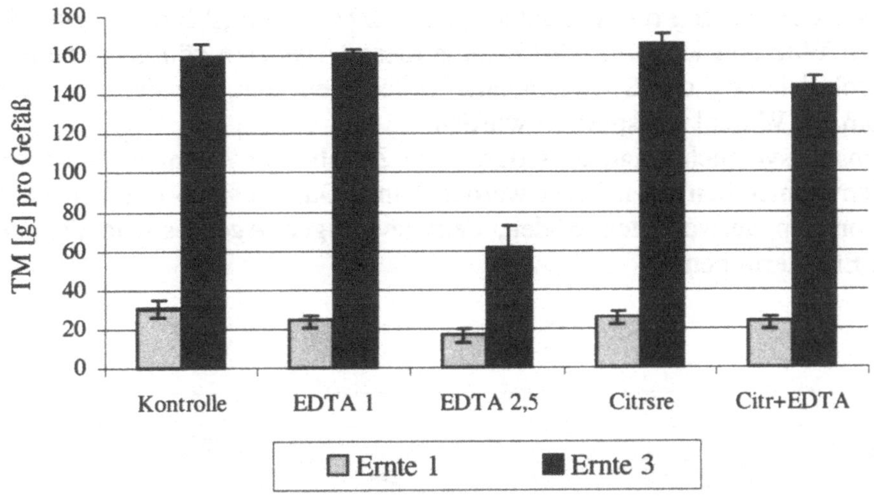

Abb. 2. Trockenmasseerträge der Sonnenblumen (in g pro Gefäß) nach unterschiedlicher Behandlung mit mobilisierenden Agenzien bei Boden 1 (Ernte 1 und 3).

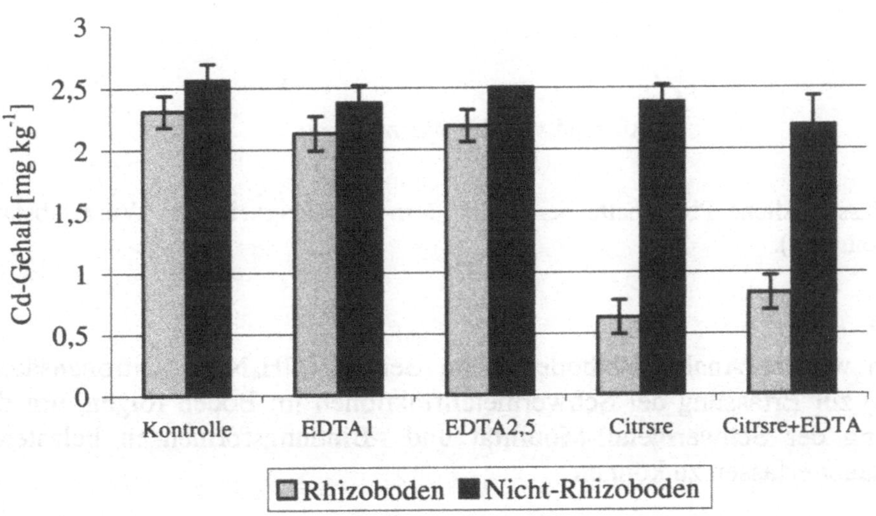

Abb. 3. Gehalte an wasserlöslichem Cd im Rhizosphärenboden und Nicht-Rhizosphärenboden nach Behandlung mit mobilisierenden Agenzien (Boden 1, Ernte 1).

Im wurzelfreien Taschenboden der Kontrollvariante (siehe Abb. 4) ist ebenfalls ein Verarmungsprofil in den ersten 5 mm Abstand von der Wurzel zu erkennen. Die stärkste Abreicherung an Pb kann in unmittelbarer Wurzelnähe beobachtet werden.
Im Gegensatz dazu zeigen die Varianten „*Citr+EDTA*“ und „*EDTA 1*“ eine Anreicherung in Wurzelnähe. Durch die Zugabe dieser Agenzien könnte es zu einer erhöhten Mobilisierung der Schwermetalle gekommen sein, die dann mit dem Massenstrom zur Wurzel transportiert wurden.
Der Gewächshausversuch zeigt, dass durch die Zugabe der Agenzien die Dynamik von Schwermetallen beeinflusst werden kann. Das Ausmaß der Effekte ist abhängig von den verwendeten Böden, dem jeweiligen Agens sowie den verschiedenen Ernteterminen.

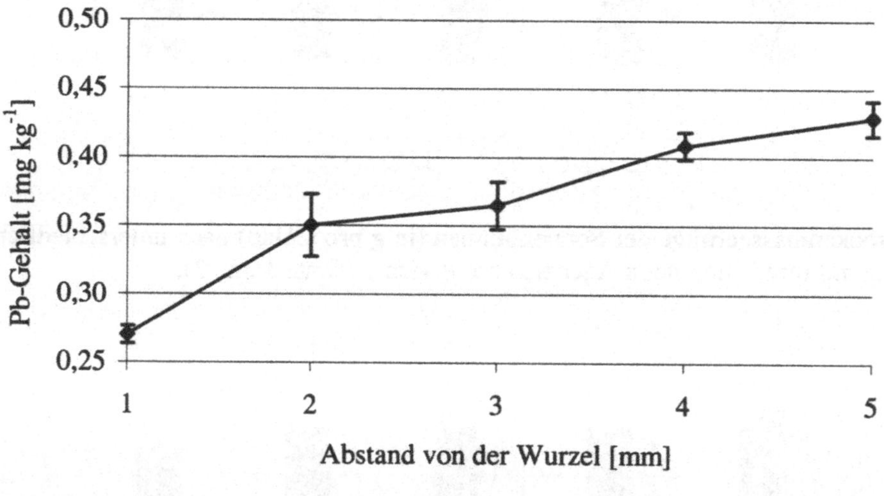

Abb. 4. Wasserlösliche Pb-Gehalte bei Boden 1 in Abhängigkeit vom Wurzelabstand (Ernte 1, Kontrolle).

Ausblick

Es werden weitere Analysemethoden (zum Beispiel NH_4NO_3-, Citronensäure-Extraktion) zur Erfassung der Schwermetallfraktionen im Boden folgen, um die Veränderung der Schwermetall-Mobilität und –Bindungsformen in belasteten Böden genauer erfassen zu können.

Danksagung
Die Autoren danken der DFG für die finanzielle Förderung des Projektes.

Literaturverzeichnis

ALEF, K., 1994: Biologische Bodensanierung – Methodenbuch. VCH Verlagsgesellschaft mbH, Weinheim, 228 S.

BBODSCHV, 1999: Bundes-Bodenschutz- und Altlastenverordnung, Bundesgesetzblatt, Teil 1, 36, 1554-1583.

HENDRIKS, L.; JUNGK, A., 1981: Erfassung der Mineralstoffverteilung in Wurzelnähe durch getrennte Analyse von Rhizo- und Restboden. *Zeitschrift für Pflanzenernährung und Bodenkunde* 144, 276-282.

HUND, K.; WIESNER, J., 1999: Zusammenfassung des Themenblock IV unter Einbeziehung der Diskussionsbeiträge zu den einzelnen Vorträgen, In: Fraunhofer-Institut für Umweltchemie und Ökotoxikologie (Hrsg.): *Pflanzenbelastung auf kontaminierten Standorten*, Berichte I/99, 218-221.

LOMBI, E.; WENZEL, W., 1999: Phytosanierung: Grundlegende Prozesse und künftige Verbesserungen. In: Fraunhofer-Institut für Umweltchemie und Ökotoxikologie (Hrsg.): *Pflanzenbelastung auf kontaminierten Standorten*, Berichte I/99, 157-158.

RILEY, D.; BARBER, S. A., 1969: Bicarbonate accumulation and pH-changes at the soybean root soil interface. *Soil Science Society of America Proceedings* 34, 905-908.

WU, J.; HSU, F. C.; CUNNINGHAM, S. D., 1999: Chelate-assisted Pb phytoextraction: Pb availability, uptake, and translocation constraints. *Environmental Science and Technology* 33, 1898-1904.

Prozessregulation in der Rhizosphäre
13. Borkheider Seminar zur Ökophysiologie des Wurzelraumes
Hrsg.: W. Merbach, B. W. Hütsch, L. Wittenmayer, J. Augustin
B. G. Teubner Stuttgart, Leipzig, Wiesbaden (2003), S. 90-96

Einfluss der CO_2-Konzentration auf den N-Transfer in Luzerne-Gras-Mischkulturen

Heidrun BESCHOW und Wolfgang MERBACH
Institut für Bodenkunde und Pflanzenernährung, Martin-Luther-Universität Halle-Wittenberg, Adam-Kuckhoff-Str. 17b, D-06108 Halle/Saale,
E-mail: beschow@landw.uni-halle.de

Abstract

Ecological agriculture expands in industrial countries. Therefore biological nitrogen fixation receives more attention. Especially grass species in mixed stands with legumes like alfalfa or clover play an important role for fodder production, because the dry mass and N yield of grass increase in comparison with grass pure stands. The N transfer between alfalfa and ryegrass under normal and enriched CO_2 concentrations was studied through isotope dilution at a very low level of N fertilization (5,5 mg N/pot) in a pot experiment using quartz sand as substrate. The growth of grass and non nitrogen fixing alfalfa plants were limited and the N-transfer between alfalfa and ryegrass was only 0,06 kg N ha^{-1} at 360 ppm CO_2 and 1,1 kg N ha^{-1} at 720 ppm CO_2 although the nitrogen fixation was nearly constant. Root/ shoot ratio and root mass of ryegrass increased at 720 ppm CO_2. We concluded that the improved root growth of grass plants promoted the uptake of fixed N from alfalfa plants. The duration of growth at 720 ppm CO_2 until harvest was only two weeks and therefore mainly root exudates from alfalfa plants should be the source of increased fixed N in grass plants.

Einleitung

„Ökologischer" Landbau findet in industrialisierten Ländern immer größere Verbreitung. Dadurch gewinnt auch die biologische Stickstofffixierung zunehmend an Bedeutung. Für die Futtergewinnung spielen vor allem Luzerne- und Klee-Gras-Kulturen eine große Rolle, da sich dieser Mischanbau fördernd auf die Trokkenmassebildung und den N-Ertrag von Gras im Vergleich zu Grasreinkulturen auswirkt. Als mögliche Ursachen für diese Steigerungen werden, z.B. eine verbesserte Aufnahme von oder eine geringere Konkurrenz um Boden-N sowie der Einfluß von Endomykorrhizen, diskutiert. Nach eigenen Untersuchungen scheint jedoch der N-Transfer von der Leguminose zur Nichtleguminose direkt über Wur-

zelabscheidungen und indirekt über die Mineralisierung abgestorbener Knöllchen und Wurzeln hauptsächlich verantwortlich zu sein (BESCHOW *et al.* 2000). Die Abscheidung N-haltiger Verbindungen durch die Wurzel ist abhängig vom Ontogenesestadium der Pflanze. Offenbar spielen veränderte source-sink-Beziehungen eine große Rolle. Diese Beziehungen sind durch Umweltbedingungen beeinflussbar. Es ist bekannt, dass sich bei erhöhter CO_2-Konzentration der Luft die Pflanzenarchitektur, die Aktivität der Stoffwechselfunktionen und insbesondere das Wurzel-Sproß-Verhältnis verändern.
Unsere Untersuchungen hatten das Ziel, den N-Transfer zwischen Luzerne und Gras bei normaler und erhöhter CO_2-Konzentration der Luft zu vergleichen, um die Regulation dieses Transfers besser beschreiben zu können.

Material und Methoden

Pflanzenwachstum und Ernte

Luzerne (*Medicago sativa* L.) und Deutsches Weidelgras (*Lolium perenne* L. „Livre") wurden in „Rosentöpfen“ (oberer Durchmesser 11,5 cm, Höhe 18 cm, unterer Durchmesser 7 cm) mit N-freiem Quarzsand (1,5 kg) als Substrat bei 60 % WK $_{max}$ angezogen. In unseren Untersuchungen wurden die nodulierende, aber nicht fixierende Luzernesorte „Insaranac“ im Vergleich zu der N_2-fixierenden Luzerne „Saranac“ verwendet. Die Luzerne wurde mit *Rhizobium meliloti*, Stamm 120F51 (DSM, Braunschweig) inokuliert. Zur N-Düngung wurde eine sehr geringe Dosis von 5,5 mg N/Gefäß in Form von KNO_3 mit 95 at% $^{15}N_{exc}$ gegeben, um den N-Transfer mittels Isotopenverdünnungsanalyse bei einer sehr niedrigen N-Gabe bestimmen zu können. Es existierten 3 Reinkulturen (Deutsches Weidelgras [Gras (R)]; Luzerne "Insaranac" [Insar (R)]; Luzerne "Saranac" [Sar (R)] und 2 Mischkulturen (Gras + Luzerne „Insaranac“ [Gras (M Insar), Insar (M Gras)]; Gras + Luzerne „Saranac“ [Gras (M Sar), Sar (M Gras)]) mit je 3 Wiederholungen.
Die Pflanzenanzahl pro Gefäß (0,008 m^2) betrug in Reinkultur 10 Graspflanzen und 5 Luzernepflanzen sowie in Mischkultur 7 Graspflanzen und 3 Luzernepflanzen. Die Versuchsdauer umfasste 12 Wochen. Die Anzuchtgefäße wurden 2 Wochen vor der Ernte in Küvetten mit konstanten CO_2-Konzentrationen von 360 bzw. 720 ppm gestellt. Sproß und Wurzeln wurden jeweils getrennt geerntet (Luzernewurzeln visuell separiert und alle Restwurzeln dem Gras zugerechnet) und die Trockenmasse, der C- und N-Gehalt sowie die ^{15}N-Abundanz bestimmt (vario-EL Elementaranalysator mit gekoppeltem NOI 7).

Berechnung der N-Bilanzen

Alle Daten werden in mg pro Gefäß angegeben.

Die totale N-Menge in Spross und Wurzel wurde aus dem N-Gehalt und der Trockenmasse bestimmt:

N_{total} = TS [mg] · N [%]/100 (1)

Der aufgenommene Düngerstickstoff wurde auf Basis von 95 at% $^{15}N_{exc}$ berechnet: $N_{Dünger}$ = mg at%$^{15}N_{exc.}$ in der Pflanze · 1,05 (2)

Bei der Ermittlung der N-Deposition wurde die Variante der nichtfixierenden Luzerne zugrunde gelegt und von N_{total} (1) der N im Samen und der aufgenommene Dünger-N (2) abgezogen:

$N_{Deposition} = N_{total} - (N_{Samen} + N_{Dünger})$ (3)

Der fixierte N in Luzerne und der N-Transfer von fixiertem N zu Gras wurde aus der Differenz der totalen Gesamtmenge und der Summe von N aus dem Samen, dem Dünger und der Deposition) kalkuliert:

Luzerne: $N_{fix} = N_{total} - (N_{Dünger} + N_{Samen} + N_{Deposition})$ (4)

Gras: $N_{Transfer\,(fix)} = N_{total} - (N_{Dünger} + N_{Samen} + N_{Deposition})$ (5)

Ergebnisse und Diskussion

Bei den Graspflanzen sind sowohl bei der Trockenmassebildung (Abb. 1) als auch bei den akkumulierten N-Mengen (Abb. 2) große Unterschiede in Abhängigkeit von der CO_2-Konzentration und dem Anbau in Rein- oder Mischkultur zu erkennen. In Mischkulturen ist bei normalem CO_2-Gehalt immer ein Abfall der Trokkenmasse und der N-Mengen in Graspflanzen im Vergleich zur Reinkultur offen-

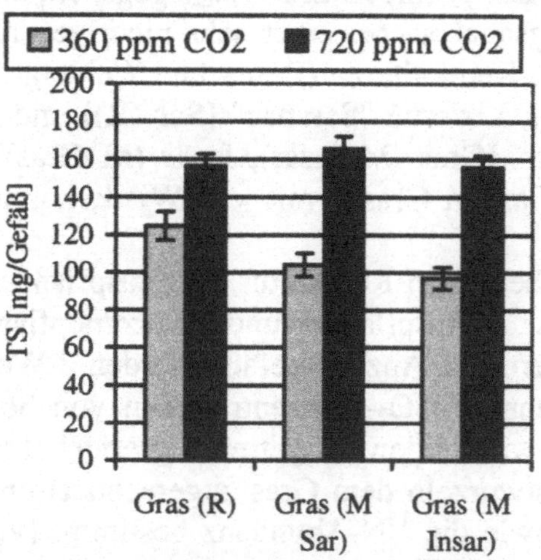

Abb. 1. Trockensubstanz von Graspflanzen in Rein- und Mischkultur mit Luzerne in mg/Gefäß, bezogen auf 10 Graspflanzen pro Gefäß.

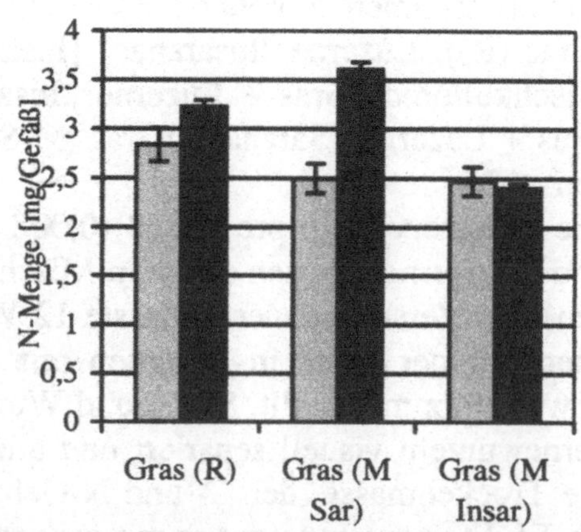

Abb. 2. N-Mengen in Graspflanzen in Rein- und Mischkultur mit Luzerne in mg/Gefäß, bezogen auf 10 Graspflanzen pro Gefäß.

bar durch die sehr geringe mineralische N-Gabe festzustellen, die auch durch N-Transfer nicht abgepuffert werden konnte. Dagegen zeichnet sich bei verdoppelter CO_2-Konzentration in Graspflanzen, angebaut mit fixierender Luzerne, bei beiden Parametern ein Anstieg gegenüber reiner Graskultur ab. Sieht man sich die N-Bilanzen an, wird deutlich, dass tatsächlich bei 720 ppm ein höherer N-Transfer zu Gras ermittelt wurde (Tab. 1), der allein 34% der gesamten N-Menge in Gras betrug. Als eine Urache für dieses Ergebnis könnte eine verbesserte N_2-Fixierung infolge einer erhöhten C-Assimilation angenommen werden. Jedoch wurde die N_2-Fixierung durch CO_2 nicht signifigant beeinflusst (Tab. 2).

Tab. 1. N-Bilanzen in Mischkulturen sowie N-Transfer von Luzerne zu Gras.

CO_2 [ppm]	N-Bilanz-größen	N [mg/Gefäß] Gras (Sar)	Gras (Insar)
360	N_{Total}	1,75	1,73
	$N_{Dünger}$	0,03	0,06
	N_{Samen}	0,20	0,20
	$N_{Deposition}$	1,47	1,47
	$N_{Transfer (fix)}$	0,05 (3%)	
720	N_{Total}	2,53*	1,68
	$N_{Dünger}$	0,03	0,06
	N_{Samen}	0,20	0,20
	$N_{Deposition}$	1,43	1,43
	$N_{Transfer (fix)}$	0,87 (34%)	

*= signifikant im Vergleich zu 360 ppm CO_2, α= 5%, t-Test

Tab. 2. N_2-Fixierung von Luzerne „Saranac" in Rein- und Mischkultur in mg N/Gefäß.

	360 ppm CO_2	720 ppm CO_2
Sar (R)	127,3	121,1^{ns}
Sar (M Gras)	99,2	81,5^{ns}

ns =nicht signifigant im Vergleich zu 360 ppm CO_2-, α= 5%, t-Test,

Die Werte für die Fixierungsleistung der Luzerne „Saranac" sinken sogar geringfügig bei 720 ppm CO_2. Es müssen also andere Gründe für den Transferzuwachs vorliegen.

Die Hauptkonsequenz von erhöhter atmosphärischer CO_2-Konzentration ist eine Steigerung der Photosyntheserate und damit eine wachsende Konzentration löslicher Zucker und Reservezucker (FARRAR und GUNN 1996). Eine dramatische Steigerung des Wurzel-Sproß-Verhältnisses (Quotient aus Trockenmasse der Wurzel und Trockenmasse des Sprosses) ist oft ein nachfolgendes Ergebnis. Dieses nichtäquivalente Wachstum von Sproß und Wurzel ist aber eher durch Defizite bei der Licht-, Wasser- und Nährstoffversorgung als durch Unterschiede in der CO_2-Assimilation zu erklären (KRAMER und BOYER 1995; CAKMAK *et al.* 1994). Forschungsbedarf besteht bei der Einbeziehung unterschiedlicher Faktoren in die C-Bilanz von Pflanzen, wie Wurzelverluste durch Absterben, Rhizodeposi-

tion, Atmung und Sproßverluste durch Verdampfen oder Leaching von Verbindungen (ROGERS *et al.* 1996). Analoges gilt unserer Meinung nach auch für die N-Bilanz.
In Tab. 3 werden die absoluten und prozentualen Veränderungen der Wurzel-Sproß-Verhältnisse in Rein- und Mischkultur bei den beiden CO_2-Stufen verglichen.

Tab. 3. Absolute Werte des Wurzel-Spross-Trockensubstanzverhältnisses und dessen prozentuale Veränderung in Rein- und Mischkulturen von Gras und Luzerne.

	360 ppm CO_2	720 ppm CO_2	% Δ (720/360 ppm)
Gras (R)	2,31	2,90	+ 26
Sar (R)	0,61	0,81	+ 33
Gras (M Sar)	2,05	2,88*	**+40**
Sar (M Gras)	0,51	0,75	**+47**

*= signifikant im Vergleich zu 360 ppm CO_2, α= 5%, t-Test

Tab. 4. Wurzelmassen von Gras- und Luzernepflanzen in Mischkultur in mg TS/Gefäß.

	360 ppm CO_2	720 ppm CO_2
Gras (M Sar)	24	**30***
Sar (M Gras)	1363	1140

*= signifikant im Vergleich zu 360 ppm CO_2,

Bei beiden Pflanzenarten ist sowohl in Rein- als auch in Mischkultur eine extreme Erhöhung des Wurzel-Sproß-Verhältnisses unter angereicherten CO_2-Bedingungen zu erkennen. In der Literatur werden Erhöhungen um 7,4% (680/340 ppm CO_2) bei Weidelgras (MORISON und GIFFORD 1984) und um 57,1% (720/360 ppm CO_2) bei Luzerne (MACDOWELL 1982) angegeben. Offenbar sollte bei unseren Versuchsbedingungen mit extremen N-Mangel eine Vergrößerung der Wurzelmasse (Tab. 4) und damit Wurzeloberfläche das Überleben der Graspflanzen durch erhöhte Aufnahmekapazität für Mineralstoffe aus dem Boden sichern.
Abbildung 3 zeigt die geernteten Versuchspflanzen. Zu erkennen ist die extreme Wurzelverlängerung der Graspflanzen bei 720 ppm CO_2.
Eine bessere N-Versorgung der Graspflanzen durch Luzerne fixiertem N_2 zeigt sich auch bei Betrachtung der C/N-Verhältnisse (Tab. 5). Beim Mischanbau mit Luzerne ist das C/N-Verhältnis in Graspflanzen enger als bei Graspflanzen in Reinkultur bei beiden CO_2-Stufen, wobei bei der hohen CO_2-Stufe der Unterschied noch stärker ausgeprägt ist und damit ebenfalls auf höheren N-Transfer hindeutet.

Tab. 5. C/N-Verhältnisse in Wurzel und Spross in Rein-und Mischkulturen von Gras und Luzerne.

	Wurzel		Spross	
	360 ppm CO_2	720 ppm CO_2	360 ppm CO_2	720 ppm
Gras (R)	12,4	**17,5**	11,7	14,9
Sar (R)	9,0	8,9	7,5	7,4
Gras (M Sar)	10,9	**13,1**	11,2	12,0
Sar (M Gras)	7,8	9,4	6,2	7,5

Abb. 3: Versuchspflanzen nach 14tägiger Anzucht in Küvetten mit 360 oder 720 ppm CO_2

Zusammenfassung

1. Das Wachstum von Deutschem Weidelgras (*Lolium perenne* L., „Livre") und nichtfixierenden Luzerne (*Medicago sativa* L., „Insaranac") wurde durch die sehr geringe N-Startdüngung limitiert und auch der N-Transfer von der fixierenden Luzerne (*Medicago sativa* L., „Saranac") zu Gras erreicht mit 0,06 kg N/ha bei 360 ppm CO_2 und 1,1 kg N/ha bei 720 ppm CO_2 nur minimale Werte.
2. Der N-Transfer stieg mit zunehmender CO_2-Konzentration der Luft, obwohl die N_2-Fixierungsleistung der Luzerne annähernd konstant blieb. Das Wur-

zel/Sproß-Verhältnis und die Wurzelmassen nahmen bei Graspflanzen zu. Die Vergößerung der Graswurzeln ermöglicht offenbar eine bessere Vernetzung mit den Luzernewurzeln und führt zu einer besseren Aufnahme von N-haltigen Verbindungen, die aus der Fixierung der Luzerne stammen.

3. Da die Begasung nur über eine relativ kurze Dauer von 2 Wochen erfolgte, wird angenommen, dass vor allem Wurzelabscheidungen am N-Transfer beteiligt gewesen sind.

Literaturverzeichnis

BESCHOW, H.; SCHULZE, J.; MERBACH, W., 2000: Transfer of symbiotically fixed nitrogen in an alfalfa-grass mixture through isotope dilution in a pot experiment. *Isotopes Environmental Health Studies* 36, 21-33.

CAKMAK, I.; HENGELER, C., MARSCHNER, H., 1994: Partitioning of shoot and root dry matter and carbohydrates in bean plants suffering from phosphorus, potassium and magnesium deficiency. *Journal of Experimental Botany* **45**, 1245-1250.

FARRAR, J. F.; GUNN, S., 1996: Effects of temperature and atmospheric carbon dioxide on source-sink relations in the context of climate change. In: E. Zamski, A. A. Schaffer (eds.) *Photoassimilate distribution in plants and crops: Source-sink relationships*. New York, USA: Marcel Dekker, 389-406.

KRAMER, P. J.; BOYER, J. S., 1995: *Water relations of plants and soils*. San Diego, CA, USA: Academic Press, 495 pp.

MACDOWELL, F. D. H., 1982: Effects of light intensity and CO_2 concentration on the kinetics of 1st month growth and nitrogen fixation of alfalfa. *Canadian Journal of Botany* 61, 731-740.

MORISON, J. I. L.; GIFFORD, R. M., 1984: Plant growth and water use with limited water supply in high CO_2 concentrations. II Plant dry weight, partitioning and water use efficiency. *Australian Journal of Plant Physiology* 11, 375-384.

ROGERS, H. H.; RUNION, G. B.; PRIOR, S. A.; TORBERT, H. A., 1996: Response of plants to elevated atmospheric CO_2: Root growth, mineral nutrition and soil carbon. In: J. R. Seemann, Y. Luo, H. A. Mooney (eds.) *Carbon dioxide and environmental stress*. San Diego, CA, USA: Academic Press, 375 pp.

Prozessregulation in der Rhizosphäre
13. Borkheider Seminar zur Ökophysiologie des Wurzelraumes
Hrsg.: W. Merbach, B. W. Hütsch, L. Wittenmayer, J. Augustin
B. G. Teubner Stuttgart, Leipzig, Wiesbaden (2003), S. 97-103

Growth and mineral nutrient acquisition of mycorrhizal soybean grown on a calcareous soil under drought stress conditions

Henry BARUS[1] and Eckhard GEORGE[2]
[1] Institute of Plant Nutrition (330), Hohenheim University; D-79593 Stuttgart, Germany
[2] Institute of Plant Production Sciences, Humboldt University Berlin, and Institute of Vegetable and Ornamental Crops (IGZ); D-14979 Großbeeren, Germany

Abstract

Plants grown on calcareous soils under drought conditions often have reduced productivity because of limited availability of mineral nutrients and water in soil. The objective of this research was to study the effect of mycorrhizal colonization on the growth and nutrient acquisition of soybean (*Glycine max* (L.) Merill cv. Marble-Arrow) plants under drought conditions. The study tested specifically the hypothesis that arbuscular mycorrhizal fungi can increase plant nutrient uptake even at a relatively high supply of soil P, and that this effect is relatively greater in dry soil than in well watered soil. At harvest, plants grown under drought conditions had decreased total biomass compared to well watered plants. Arbuscular mycorrhizal (AM) plants under drought conditions had significantly higher P and Zn concentrations and contents in shoots compared to non-mycorrhizal (nonAM) plants. In contrast, shoot P concentrations and total P content were not increased in AM plants under well watered conditions. Improved drought tolerance of AM soybean plants can be due to increased uptake of immobile ions such as P and Zn.

Introduction

Many previous studies have shown that AM fungi under well watered conditions can increase the nutrient uptake capacity of host plants. This is particularly true for ions with reduced mobility in soils, such as P, Zn or Cu. In contrast, results on a direct effect of mycorrhizal colonisation on plant water uptake and water use are inconsistent. Some authors report that AM fungi help their host plants to take up water (FABER *et al.* 1991, RUIZ-LOZANO and AZCÓN 1995) while others reported no effect (GEORGE *et al* 1995, BRYLA and DUNIWAY 1997). AM plants may benefit from increased hyphal extraction of soil water and transfer to the host plant (FABER *et al.* 1991). However, GEORGE *et al.* (1992) found no evidence of significant water transport by AM hyphae, and also BRYLA and DUNIWAY (1997) reported that the ability of AM roots to extract water from soil during extreme drought was not different from that of nonAM roots.

Plants can limit the effects of drought stress on growth and function by different means, for example by increased nutrient uptake. Investigations on the effect of mycorrhiza on host plant growth and nutrient uptake under drought conditions have been documented previously (FITTER 1988). Also, mycorrhizal fungi can increase root length density or alter root system morphology (KOTHARI *et al.* 1990a).

The objective of the present research was to evaluate whether the arbuscular mycorrhizal fungus *Glomus mosseae* BEG 107 can improve growth and nutrient acquisition of soybean plants, particularly when soil moisture is limiting. The hypothesis was that AM fungi could not improve plant growth and nutrient acquisition when water and nutrients were freely available in soil, but may improve plant growth under water-limiting conditions by an increase in uptake of immobile nutrients.

Materials and Methods

This experiment was conducted in a greenhouse in Stuttgart (Germany) at average day/night temperatures of 28/15°C during summer season. Soybean (*Glycine max* L. Merr. cv. Marble-Arrow) plants were grown in a calcareous soil (Luvisol; subsoil of a loess). The inoculum used was *Glomus mosseae* (BEG 107). Basal nutrients were applied in the following concentration and form : 300 mg N (NH_4NO_3), 200 mg K (KCl), 100 mg Mg ($MgSO_4.7H_2O$), 2 mg Fe (NH_4-Fe-citrate, 28 % Fe), 10 mg Zn ($ZnSO_4.7H_2O$), 10 mg Cu ($CuSO_4.5H_2O$) per kg soil. Phosphorus was applied at the relatively high rate of 150 mg P (KH_2PO_4) per kg soil to all treatments, to minimise overall growth differences between AM and nonAM plants. The soil was mixed with 10 % inoculum and each pot contained 4.0 kg soil. NonAM plants received a similar amount of inoculum sterilised by autoclaving (120°C, 20 min) plus a filtrate of non-sterilised mycorrhizal inoculum that was sieved through a 30-μm filter, in attempt to provide similar microbial populations like in the original inoculum for AM plants, except AM fungi. The seeds were sterilised with 10 % H_2O_2 for 2 min and then pre-germinated for eight hours in saturated $CaSO_4$ and sown directly into the pot.

Plants were watered daily until drought-stress treatments were initiated 45 days after sowing. Drought stress was imposed by withholding water from pots until a soil water content of 6.5 % (w/w) was obtained. A time domain reflectometry (TDR; Tektronix 1502 C, Oregon) instrument was used to measure the soil water content in soil. Pots with well-watered plants were maintained at a soil water content of 21 % (w/w). Drought-stress was imposed for two weeks during pod forming stage. Thereafter, pots were re-watered and kept at full irrigation for another two weeks during pod filling/set stage. Plants were harvested 72 days after planting by severing shoots from roots. Roots were washed from soil, and root samples were taken for determination of mycorrhizal colonisation and root length, and analysed according to KOSKE and GEMMA (1989) and GIOVANNETTI and MOSSE (1980). In dried and ground shoot and root tissue, after digestion P was measured with the spectrophotometric phosphomolybdate method, K and Ca by flame photometry, and Mg, Zn, Cu, Mn, and Fe by atomic absorption spectroscopy.

Results and Discussion

Plants grew well and no symptoms of nutrient deficiency were observed. Inoculated plants were highly colonised (Tab. 1). No AM colonisation was observed in nonAM plants. Shoot and root biomass were not significantly different in AM plants and nonAM plants, but drought conditions decreased shoot growth (Tab. 1). AMF colonisation also did not affect the root-shoot ratio, but drought stress increased the root-shoot ratio.

Tab. 1. Colonisation rate, biomass, and root/shoot ratio in soybean plants. Two Way ANOVA followed by the Fisher LSD test for mean separation was used to test for statistical differences. Different letters in each column indicate that there is a significant difference among means.

	Colonisation rate (%)	Biomass (g) Shoot	Root	Total	Root/Shoot ratio
Wet					
AM+	70	23.5a	1.5	25.0a	0.06a
AM-	0	25.0a	1.8	26.8a	0.07a
Dry					
AM+	58	15.7b	1.5	17.1b	0.09b
AM-	0	13.5b	1.3	14.8b	0.10b
LSD (0.05)		2.4	-	2.6	0.01
H_2O	ns	***	ns	***	**
AM	-	ns	ns	ns	ns
H_2O x AM	-	ns	ns	ns	ns

Drought stress decreased the concentrations of P and Cu in shoots, and increased Mg and Ca concentrations (Fig. 1a-b). Due to decreased growth, under drought stress the total nutrient content in shoots was decreased for all elements (Fig. 1c-d). For concentrations in shoot or roots, in several elements we observed a significant interaction for the experimental factors water supply and mycorrhizal colonisation. For example, under drought conditions mycorrhizal colonisation increased the shoot concentrations of P and Zn, and decreased the shoot concentrations of Mn and Fe. In well watered soil, shoot P concentrations were not significantly affected by mycorrhizal colonisation. Under drought, root concentrations of P, Ca and Cu were increased, and root concentrations of Mg and Mn were decreased in AM plants compared to nonAM plants (Fig. 1a-b). Total shoot content of P and Zn was significantly increased by mycorrhizal colonisation under drought conditions, while total shoot content of other elements was not significantly affected by mycorrhizal colonisation under these conditions (Fig. 1c-d).

Phosphorus was supplied in this experiment at a relatively high level in soil. In this situation, as expected, mycorrhizal colonisation did not significantly affect total plant biomass (Tab. 1) or total plant P uptake (Fig. 1c) when plants were well-watered, probably because plant growth was not limited by P. For the same reason, P concentrations in the shoot were not significantly different between AM and nonAM plants under well-watered conditions. Under drought conditions, the increased root-shoot ratio indicated an adaptation of plants to reduced soil water availability. Mycorrhizal colonisation did not affect this adaptation significantly. Under drought conditions, however, AM plants had increased concentrations and contents in shoot and roots compared to nonAM plants of the soil-immobile elements P and Zn.

These results confirm that AM plants can have enhanced nutrient uptake, in particular of immobile ions such as P, Zn and Cu (KOTHARI *et al.* 1990b, LI *et al.* 1991). Usually, this is observed on low-P soils. Here, we show that under drought situations also on soils with high P fertilisation AM fungi can be important for the plant in uptake of P and Zn. Drought stress increased concentration of Ca and Mg in shoots, but decreased concentrations of P and Cu (Fig. 1a-b). Thus, under drought stress mycorrhizal colonisation specifically increased the uptake of nutrients that are difficult to acquire for the plants under these conditions. Therefore, the AMF association can improve nutrient uptake and thus plant tolerance under drought conditions.

The present results also confirm earlier observations (KOTHARI *et al.* 1991) that Mn and Fe concentrations can be decreased in shoots of AM plants. Mycorrhizal fungi may decrease high Mn availability in soils by both chemical and microbial alterations in the rhizosphere (POSTA *et al.* 1994).

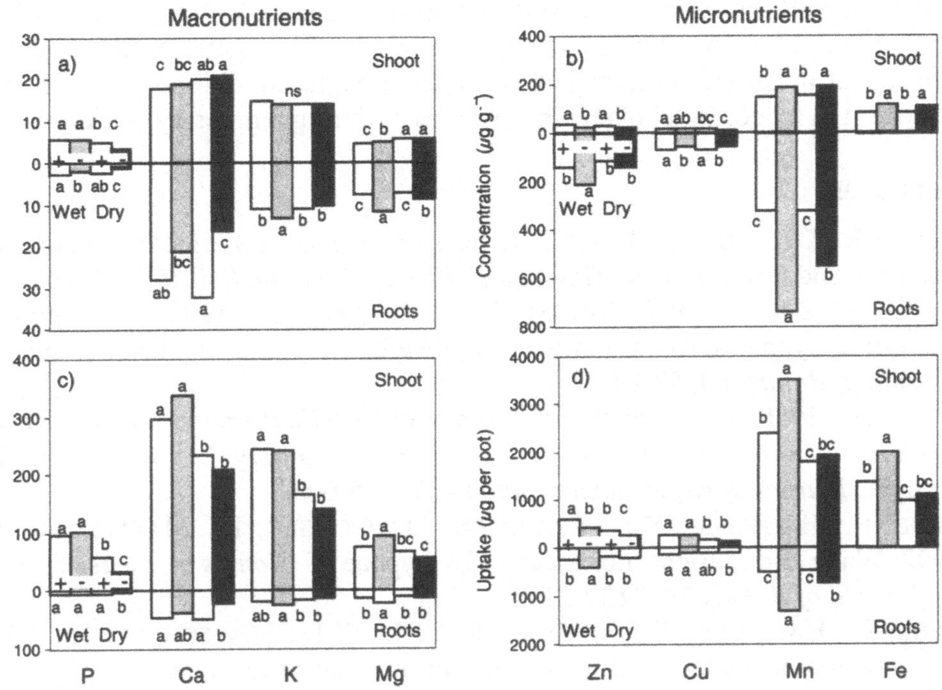

Fig. 1(a-d). Concentrations and contents of macronutrients and micronutrients in shoot and roots of soybean plants as affected by colonisation with arbuscular mycorrhiza fungi (AMF) (dark coloum = with AMF; white coloum = without AMF) and water treatment (wet = well-watered; dry = drought stress). The data were analysed with a two way ANOVA. Columns for each element marked by the same letter are not significantly ($p<0.05$) different by the Fisher LSD-test.

In a study of metal distribution in intracellular hyphae and arbuscules of an AM fungus in the root cortex of a cucumber plant, LEE (2002) showed by X-ray microanalysis that Cu was localised mainly in cortical root cell compartments, and in addition that fungal structures were Cu-enriched at low P supply. Zinc, however, was located more homogenously in different fungal structures of cortical cells (especially arbuscular cytoplasm and poly-P granules). When a higher level of P was supplied, Zn (but not Cu) was mobilised and transported to the shoot. In the present study, P was also applied at a quite high level (150 mg P/kg soil). This may explain why mycorrhizal colonisation in the present experiment decreased Zn concentrations in the roots, but increased Cu concentrations in the roots (Fig. 1b). The results confirmed that AMF plays no role in plant nutrient acquisition when nutrients, especially P, are freely available in a soil for the plant. However, when water is a limiting factor, AM fungi can improve nutrient acquisition not only on low-nutrient soils, but also on nutrient-rich soils (this study). We did not find any indication that mycorrhizal colonisation can directly affect water uptake or use by the plant, but increased nutrient uptake under drought conditions may convey a large benefit to mycorrhizal plants in dry areas or during temporary soil dryness.

References

BRYLA, D.R.; DUNIWAY, J.M., 1997: Effects of mycorrhizal infection on drought tolerance and recovery in safflower and wheat. *Plant and Soil* 197, 95-103.

FABER, B.A.; ZASOSKI, R.J.; MUNNS, D.N.; SHACKEL, K., 1991: A method for measuring hyphal nutrient and water uptake by mycorrhizal plants. *Canadian Journal of Botany* 69, 87-94.

FITTER, A.H., 1988: Water relation of red clover *Trifolium pratense* L. as affected by VA mycorrhizal infection and phosphorus supply before and during drought. *Journal of Experimental Botany* 39, 595-603.

GEORGE, E.; HÄUSSLER, K.; VETTERLEIN, D.; GORGUS, E.; MARSCHNER, H., 1992: Water and nutrient translocation by hyphae of *Glomus mosseae*. *Canadian Journal of Botany* 70, 2130-2137.

GEORGE, E.; HÄUSSLER, K.; KOTHARI, S.K.; LI, X.L.; MARSCHNER, H., 1995: Contribution of mycorrhizal hyphae to nutrient and water uptake of plants. In: Read, Lewis, Fitter and Alexander (eds) *Mycorrhizas in ecosystems*, 42-47.

GIOVANNETI, M.; MOSSE, B., 1980: An evaluation of techniques for measuring vesicular-arbuscular mycorrhizal infection in roots. *New Phytologist* 84, 489-500.

KOSKE, R.E.; GEMMA, J.N., 1989: A modified procedure for staining roots to detect VA mycorrhizas. *Mycology Research*. 92, 486-505.

KOTHARI, S.K.; MARSCHNER, H.; GEORGE, E., 1990a: Effect of VA mycorrhizal fungi and rhizosphere microorganisms on root and shoot morphology, growth and water relations in maize. *New Phytologist* 116:303-311.

KOTHARI, S.K.; MARSCHNER, H.; RÖMHELD, V., 1990b: Contribution of VA mycorrhizal fungi in acquisition of phosphorus and zinc by maize grown in calcareous soil. *Plant and Soil* 131, 177-185.

KOTHARI, S.K.; MARSCHNER, H.; RÖMHELD, V., 1991: Effect of a vesicular-arbuscular mycorrhizal fungus and rhizosphere micro-organisms on manganese reduction in the rhizosphere and manganese concentrations in maize (*Zea mays* L.). *New Phytologist*. 117, 649-655.

LEE, Y.J., 2002: Contribution of arbuscular mycorrhizal fungi to metal cations uptake in vegetables. Ph.D dissertation, University of Hohenheim, Stuttgart, Germany

LI, X.L.; MARSCHNER, H.; GEORGE, E., 1991: Acquisition of phosphorus and copper by VA-mycorrhizal hyphae and root-to-shoot transport in white clover. *Plant and Soil* 136, 49-57.

POSTA, K.; MARSCHNER, H.; RÖMHELD, V., 1994: Manganese reduction in the rhizophere of mycorrhizal and non mycorrhizal maize. *Mycorrhiza* 5, 119-124.

RUIZ-LOZANO, J.M.; AZCÓN, R., 1995: Hyphal contribution to water uptake in mycorrhizal plants as affected by the fungal species and water status. *Physiologia Plantarum* 95, 472-478.

4

Wirkungen wurzelbürtiger Verbindungen

Prozessregulation in der Rhizosphäre
13. Borkheider Seminar zur Ökophysiologie des Wurzelraumes
Hrsg.: W. Merbach, B. W. Hütsch, L. Wittenmayer, J. Augustin
B. G. Teubner Stuttgart, Leipzig, Wiesbaden (2003), S. 107-114

Eine Kultivierungsmethode zur Untersuchung von Phytohormoneffekten auf Rhizosphärenprozesse

Lutz WITTENMAYER, Annette DEUBEL, Judith WOLLNY und Kristin TRIEBSWETTER
Institut für Bodenkunde und Pflanzenernährung, Martin-Luther-Universität Halle-Wittenberg, Adam-Kuckhoff-Straße 17b, D-06108 Halle/Saale,
e-mail: wittenmayer@landw.uni-halle.de

Abstract

Higher plants affect processes in the rhizosphere by root exudation. The exuded compounds may directly influence nutrient availability or may serve as a substrate for various nutrient-solubilizing soil microorganisms. Furthermore, plant growth promoting effects of rhizosphere microbes include the production of a vast range of metabolites (biological active substances) that may affect plant growth directly after being taken up by the plant, or indirectly by modifying the soil environment. Phytohormones like auxines (indole-3-acetic acid, IAA), gibberellins (gibberellic acid, GA_3) or cytokinins (*trans*-zeatin, *t*-Z) belong to this groups of substances. It is well documented that cultures of several plant growth promoting microorganisms produce these phytohormones in considerable amounts. But there are only few data available about long-term effect of these coumpounds on rhizosphere processes of intact plants.

In the present investigation, four cultivars of maize plants (*Zea mays* L.) were grown in a greenhouse pot experiment with coarse sand as substrate. Control plants were supplied with nutrients by daily watering with a nutrient solution (40 mg P/l) and P-deficient plants were cultivated in a substrate containing 324 mg of sparingly soluble $Ca_3(PO_4)_2$ (65 mg P). Additionaly, the concentration of IAA, GA_3 or *t*-Z in the rhizosphere of maize plants was maintained at 0.1 mM (IAA, GA_3) or 0.01 mM (*t*-Z) by daily addition of a freshly mixed hormone-containing nutrient solution to one half of each P treatment. Three factorial analysis of variance showed significant effects of hormone treatment, P availability, and genotype on growth of maize plants indicating that the presented method is suitable for induction and investigation of long-term hormone effects on rhizosphere processes of higher plants.

Einleitung

Rhizosphärenprozesse zeichnen sich durch komplexe Wechselwirkungen zwischen höherer Pflanze, Mikroorganismen und Boden aus. Der Einfluss der Pflanzen über quantitativ bedeutsame Wurzelabscheidungen, wie Kohlenhydrate, Carboxysäuren, Aminosäuren, Phenole, Phytosiderophore oder Mucilage, auf biochemische Abläufe im durchwurzelten Bodenraum ist Gegenstand zahlreicher Untersuchungen (FRANKENBERGER und ARSHAD, 1995; SCHILLING *et al.*, 1998). Die genannten Verbindungen können beispielsweise unmittelbar oder mittels Mikroben den Gehalt an pflanzenverfügbaren Nährstoffen im Boden beeinflussen. Hingegen ist die Wirkung von mikrobenbürtigen Phytohormonen auf diese Prozesse bisher nur wenig untersucht worden. Zwar gelang bei isolierten Stämmen wachstumfördernder Mikroorganismen, die auf künstlichen Nährmedien wuchsen, der Nachweis der Fähigkeit zur Biosynthese von Phytohormonen, jedoch gibt es kaum Erkenntnisse über die Langzeitwirkungen dieser in der Rhizosphäre freigesetzten Spurenverbindungen auf höhere Pflanzen. Eine methodische Schwierigkeit bei der Untersuchung dieser Fragestellung besteht in der kurzen Halbwertszeit der Verbindungen und damit in der Aufrechterhaltung einer bestimmten Hormonkonzentration im Wurzelbereich über einen langen Zeitraum. Im folgenden wird ein Anzuchtverfahren vorgestellt, das die Untersuchung der Langzeitwirkung von Phytohormonen in der Rhizosphäre höherer Pflanzen erlaubt. Die vorgestellte Methode soll dazu dienen, die Wirkung von wachstumsfördernden Phytohormonen, wie Gibberelline, Auxine und Cytokinine auf morphologische und physiologische Prozesse zu untersuchen.

Material und Methoden

Als Versuchspflanzen dienten die vier verschiedenen Maissorten ‚Larex', ‚Lenz', ‚Boss' und ‚Bezemara'. Das auf ein einheitliches, sortenspezifisches Gewicht kalibrierte Saatgut wurde zunächst in Keimschalen in Sand mit $CaSO_4$-gesättigter Lösung vorgekeimt (vier Tage bei 22°C). Nachdem die Keimwurzeln eine Länge von ca. 1 cm erreicht hatten, gelangten die Sämlinge durch Umsetzung in „Rosentöpfe". Die Anzuchtgefäße hatten einen oberen Durchmesser von 11,5 cm, eine Höhe von 18 cm und verjüngten sich nach unten auf einen Durchmesser von 7 cm. Die Behältnisse besaßen an den Seitenflächen vier 6 mm breite und 5,5 cm hohe Spalte und im Boden drei Löcher mit einen Durchmesser von 7 mm. Um die Seitenöffnungen zu verschließen und um eine bessere mechanische Stabilität der Plastikgefäße zu erreichen, wurden jeweils zwei ineinandergesteckte Rosentöpfe verwendet. Der Innenraum wurde zusätzlich mit einer am Boden geschlitzten Folientüte (18 × 25 cm) ausgekleidet, die bei der Ernte eine vollständige Entnahme

des Anzuchtsubstrates und eine schonende Freilegung des Wurzelballens aus dem Anzuchtgefäß erlaubte.
Das Anzuchtgefäß nahm 1500 g handelsüblichen gewaschenen Aquarienkies (Quarzsand mit einer Körnung 1 bis 2 mm, Fa. Reißig) und maximal 110 ml Nährlösung auf. Die verwendete Rezeptur entsprach weitestgehend der von SMITH *et al.* (1983) für Maispflanzen optimierten Ruakura-Nährlösung. Anstatt des in der Originalvorschrift enthaltenen Eisen(III)-Citrates wurde eine äquimolare Menge an Eisen(III)-chlorid eingesetzt, um Überlagerungseffekte mit der Citratexsudation der Maiswurzel auszuschließen. Im Einzelnen erfolgte die Fertigung der Nährlösung wie folgt: Zunächst war die Herstellung von vier Stammlösungen erforderlich, aus denen die Gebrauchslösungen gemischt wurden:

Tab.1. Elementkonzentrationen in der Komplett- und P-freien Nährstofflösung

Element	P-Variante	
	+P	-P
N	261,6	261,6
P	40,5	0
K	241,0	241,0
S	60,9	88,6
Ca	128,2	128,2
Mg	21,1	21,1
Cu	0,0406	0,0406
Zn	0,253	0,253
Mn	0,507	0,507
Fe	10,1	10,1
B	0,507	0,507
Mo	0,0102	0,0102
Cl	29,3	29,3
Na	14,6	14,6

1 Liter Stammlösung *A*:
4,94 g $Mg(NO_3)_2$ x 6 H_2O,
16,78 g $Ca(NO_3)_2$ x 4 H_2O,
8,48 g NH_4NO_3,
2,28 g KNO_3;

1 Liter Stammlösung *B*:
2,97 g KH_2PO_4,
2,14 g K_2HPO_4 x 3 H_2O,
6,62 g K_2SO_4,
0,60 g Na_2SO_4,
0,33 g NaCl;

1 Liter Stammlösung *C*:
128,8 mg H_3BO_3,
4,84 mg $CuCl_2$ x 2 H_2O,
81,10 mg $MnCl_2$ x 4 H_2O,
0,83 mg $(NH_4)_6Mo_7O_{24}$ x 4 H_2O,
23,45 mg $ZnCl_2$,
2,18 g $FeCl_3$ x 4 H_2O;

1 Liter Stammlösung *D*:
9,97 g K_2SO_4,
60 g Na_2SO_4,
0,33 g NaCl.

Zur Herstellung der Gebrauchslösung für die Vollernährungsvarianten wurden 90 ml Lösung *A*, 90 ml Lösung *B* und 45 ml Lösung *C* in ca. 1 Liter deionisertem

Wasser gemischt, auf zwei Liter aufgefüllt und auf *p*H 6,5 mit 0,1 N Natronlauge oder 0,1 N Salzsäurelösung eingestellt.
Für die Phosphat-Mangelvarianten wurde eine P-freie Gebrauchslösung eingesetzt, bei der anstatt Lösung *B* die Stammlösung *D* verwendet wurde. Tab. 1 fasst die Elementkonzentrationen in der Komplett- und P-freien Nährstofflösung zusammen. Als alleinige Phosphatquelle für die P-Mangelpflanzen diente schwerlösliches Tricalciumphosphat $Ca_3(PO_4)_2$, das in einer Menge von 324 mg dem noch trockenen Anzuchtsubstrat bei Versuchsansatz zugefügt wurde. Die eingesetzte Salzmenge entspricht 65 mg Phosphor. Lineares Schütteln von Tricalciumphosphat mit dem Aquarienkies reichte für eine gleichmäßige Vermischung nicht aus. Zur Herstellung eines homogenen Substrates war es erforderlich, beide Komponenten in einen 5-l-Erlenmeyer-Kolben zu überführen und diesen mindestens zwei Minuten um seine Längsachse zu drehen. Danach konnte das Substrat in die Rosentöpfe umgefüllt werden.
Die einzelnen Phytohormone wurden zunächst in 96%igem Ethanol gelöst: 219 mg 3-Indolylessigsäure (IES), 433 mg Gibberellinsäure A_3 (GA_3) in jeweils 5 ml und 25 mg *trans*-Zeatin (*t*-Z) in 4,5 ml. Diese Angabe für das jeweilige Hormon entspricht ungefähr den im angegebenen Lösungsmittelvolumen maximal lösbaren Substanzmengen. Eine hohe Hormonkonzentration der Stammlösung wurde angestrebt, um negative Nebenwirkungen des Lösungsmittels auf die Maispflanzen zu vermeiden. 200 μl der Hormonstammlösungen ergaben nach Zugabe zu 500 ml Nährstoffgebrauchslösung eine Hormonkonzentration von 0,1 mM (IES, GA_3) oder 0,01 mM (*t*-Z). Die Hormonzugabe geschah unmittelbar vor der Nährstoffverabreichung. Die hormonfreien Nährlösungen erhielten zum Vergleich 200 μl Ethanol.
Die Menge der verabreichten Nährstofflösungen richtete sich nach der Größe der Maispflanzen und damit nach deren Wasserverbrauch. Das applizierte Volumen sollte unabhängig von Sorte, Nährstoffregime oder Hormonbehandlung sein. Eventuelle sortenspezifische Unterschiede in der Transpiration wurden deshalb durch zusätzliche Gaben von deionisiertem Wasser ausgeglichen, so dass für die Sorten auf gleicher P-Stufe vergleichbare Hormon- und Nährstoffmengen eingesetzt wurden. Abbildung 1 zeigt den kumulativen Verbrauch an Nährstofflösung im Versuchsverlauf.
Bis zur Ernte verbrauchten die Pflanzen insgesamt 1620 ml Nährlösung, was in der +P-Variante einer P-Gabe von 65 mg entspricht. Diese P-Menge ist somit auch mit der in Form von Tricalciumphosphat verabreichten P-Menge vergleichbar.
Sechs Wochen nach dem Umpflanzen der Keimlinge erfolgte die Ernte. Da zu diesem Entwicklungsstadium die Fahne noch nicht sichtbar war, wurden zur Er-

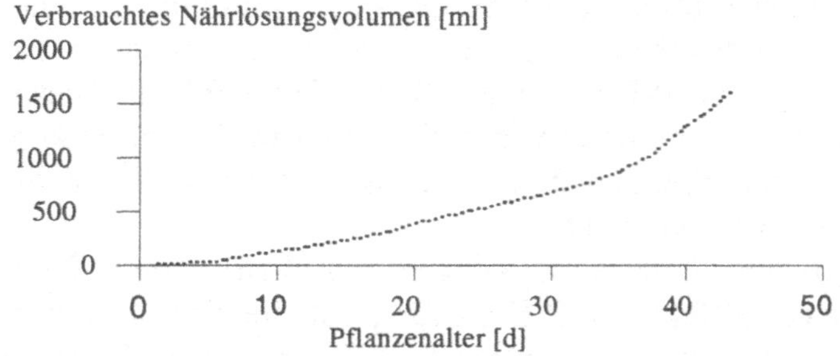

Abb. 1. Kumulativer Verbrauch der Nährstofflösung in Abhängigkeit vom Pflanzenalter.

fassung der Sprosslänge die Bätter am Spross nach oben gerichtet und der Abstand zwischen Wurzelhals und der höchsten Blattspitze mit einem Lineal erfaßt. Der Durchmesser des Stengels wurde in Höhe des ersten echten Laubblattes gemessen.
Zur Beurteilung der Effekte der Prüffaktoren auf die ausgewählten Prüfmerkmale wurde eine dreifakorielle Varianzanalyse gerechnet (BÄTZ *et al.* 1982, S. 140 ff).

Ergebnisse und Diskussion

Die F-Tests aus der Varianzanalyse für die Prüfmerkmale Sprosslänge, -durchmesser sowie -frischmasse sind in Tab. 2 zusammengefaßt.

Tab. 2. F-Test für die Prüfmerkmale Sprosslänge, -durchmesser sowie -frischmasse

Prüffaktor	Spross-Prüfmerkmal			F-Wert	
	Länge	Durch-messer	Frisch-masse	P = 0,05	P = 0,01
A (Hormon-behandlung)	352,2	45,9	277,0	2,7	4,0
B (P-Ernährung)	20,3	56,3	743,4	3,9	6,9
C (Sorte)	13,5	2,4	26,1	2,7	4,0
A × B	0,3	1,9	8,5	2,7	4,0
A × C	1,3	1,8	6,6	2,0	2,6
B × C	6,1	1,1	6,2	1,9	4,0
A × B × C	0,6	2,3	2,2	2,0	2,6

Für die untersuchten Prüfmerkmale Sprosslänge, -durchmesser und -frischmasse konnte eine signifikante bis hochsignifkante Beeinflussung durch die Prüffaktoren Hormonbehandlung (A), P-Ernährung (B) und Sorte (C) nachgewiesen werden. Lediglich der Sprossdurchmesser war nicht sortenspezifisch. Es konnte somit gezeigt werden, dass alle drei gewählten Prüffaktoren dauerhafte Veränderungen im Pflanzenwachstum bewirkt haben.
Aufgrund signifikanter Wechselwirkungen zwischen einigen Prüffaktoren, beispielsweise war der Effekt der P-Ernährung auf die Sprosslänge von der Sorte abhängig (B × C signifikant), ist in solchen Fällen nur ein Vergleich von ABC-Mittelwerten auf gleicher AB-, AC- oder BC-Stufe sinnvoll. Dies soll im Folgenden für das Prüfmerkmal Sprosslänge exemplarisch gezeigt werden.
Die vier geprüften Maissorten zeigten gleichgerichtete Reaktionen auf die Applikation der verwendeten Hormone. GA_3 und *t*-Z wirkten signifikant sprossverlängernd, während bei IES-Behandlung keine wesentliche Veränderung der Sprosslänge zu beobachten war (Abb. 2 bis 5).
Bekanntlich ist die wachstumsfördernde Wirkung von Auxinen konzentrationsabhängig und kann bei Applikation oberhalb des Optimalbereiches sogar das Pflanzenwachstum hemmen (Epinastie), was bei der Unkrautbekämpfung angewendet wird. Untersuchungen mit nierdrigeren IES-Konzentrationen sind erforderlich, um die Frage nach einer optimalen Auxindosis in der Rhizosphäre von Maispflanzen zu klären.
In Bezug auf die P-Versorgung zeigten im allgemeinen die Pflanzen ohne Hormonapplikation den deutlichsten Unterschied zwischen +P- und -P-Variante. Dieser wurde durch die Hormonbehandlungen aufgehoben. Lediglich bei der Sorte ‚Bezemara' waren bei allen Hormonvarianten die -P-Pflanzen den vollernährten unterlegen (Abb. 5).
Neben den deutlichen Unterschieden im Wachstum der Maispflanzen sind weitere Aspekte der Phytohormonwirkung zu betrachten. So soll der P-Gehalt einzelner Organe, die Aktivität saurer Phosphatasen im Spross und in den Wurzelabscheidungen, die Menge an wasserlöslichem P im Substrat, die Wurzeloberfläche und die Zusammensetzung der Wurzelabscheidungen untersucht werden.

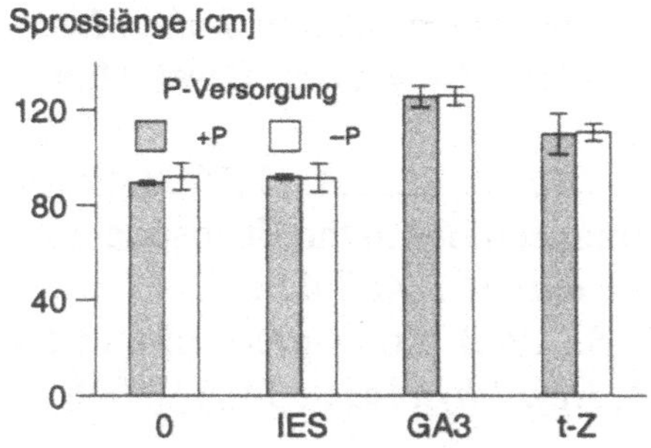

Abb. 2. Einfluss der P-Versorgung und Hormonapplikation auf die Sprosslänge bei der Maissorte ‚Larix'.

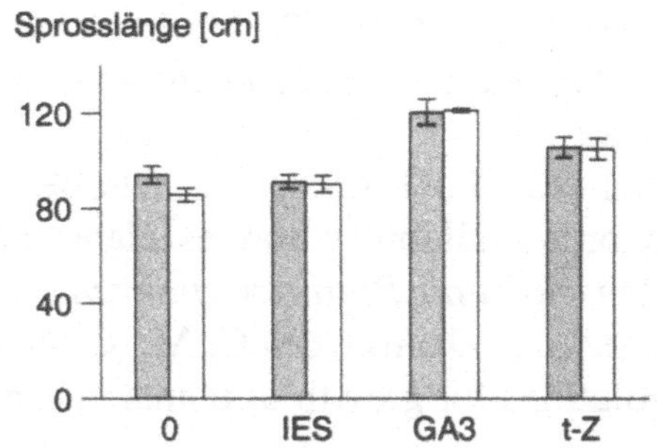

Abb. 3. Einfluss der P-Versorgung und Hormonapplikation auf die Sprosslänge bei der Maissorte ‚Lenz'.

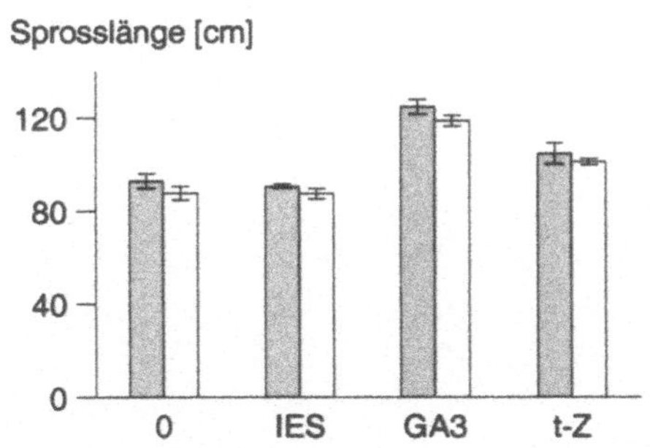

Abb. 4. Einfluss der P-Versorgung und Hormonapplikation auf die Sprosslänge bei der Maissorte ‚Boss'.

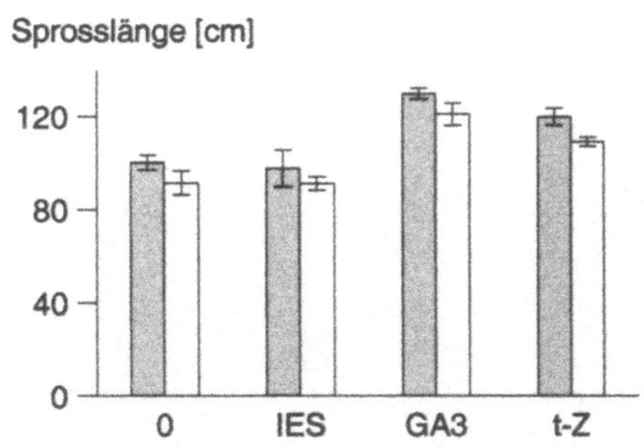

Abb. 5. Einfluss der P-Versorgung und Hormonapplikation auf die Sprosslänge bei der Maissorte ‚Bezemara'.

Danksagung

Die Autoren danken Frau E. Pfeil und Frau Chr. Birke für die Unterstützung bei der Durchführung des Versuches und Herrn Dr. R. Georg, RAGT Saaten Deutschland GmbH, Herford, für die kostenlose Bereitstellung des Saatgutes der Sorte ‚Boss'.

Literaturverzeichnis

BÄTZ, G.; DÖRFEL, H.; FUCHS, A.; THOMAS, E., 1982: *Einführung in die Methodik des Feldversuchs.* VEB Deutscher Landwirtschaftsverlag, Berlin.

FRANKENBERGER, W. T., Jr.; ARSHAD, M., 1995: *Phytohormones in Soils: Microbial Production and Function* . Marcell Dekker, Inc. New York, Basel, Honkong

SCHILLING; G.; GRANSEE, A.; DEUBEL, A.; LEŽOVIČ, G.; RUPPEL, S.,1998: Phosphorus availability, root exudates, and microbial activity in the rhizosphere. *Zeitschrift für Pflanzenernährung und Bodenkunde* 161, 465-478.

SMITH, G. S.; JOHNSTON, C. M.; CORHFORTH, I. S., 1983: Comparison of nutrient solutions for growth of plants in sand culture. *New Phytologist* 94, 537-548

Prozessregulation in der Rhizosphäre
13. Borkheider Seminar zur Ökophysiologie des Wurzelraumes
Hrsg.: W. Merbach, B. W. Hütsch, L. Wittenmayer, J. Augustin
B. G. Teubner Stuttgart, Leipzig, Wiesbaden (2003), S. 115-122

Phosphatmobilisierende Wirkung verschiedener wurzel- bzw. mikrobenbürtiger organischer Kohlenstoffverbindungen in unterschiedlichen Böden

Annette DEUBEL und Wolfgang MERBACH
Institut für Bodenkunde und Pflanzenernährung der Martin-Luther-Universität Halle-Wittenberg, Adam-Kuckhoff-Straße 17b, D-06108 Halle / Saale

Abstract

Root derived organic carbon compounds and their microbial metabolites are of ecological importance for the nutrient supply of higher plants. An extraction method similar to the double-lactate (DL) method (2 g soil, 100 ml solution with 5 g l^{-1} substance, pH 3.7, 90 min) was used to measure the P-solubilizing ability of sugars and carboxylic acids in 10 different soils. Neutral extraction solutions were used to detect specific anion effects. Additionally a 24 h extraction at 28 °C was used to permit chemical and microbial changes of the substances in the soil.
Sugars and sugar alcohols can contribute to P mobilization in insterile soils with a pH below 7. They have no effect under alkaline conditions. The acidic extraction with gluconic acid was as efficient as DL, although the buffer capacity of the solution is relatively low. The P release increased during 24 h. However, a neutral gluconate extraction was only comparable with sugar extraction. Succinic acid was only effective under acidic conditions.
Citrate and oxalate, even in small concentrations, mobilized large P amounts under different soil conditions. Mainly the effectiveness of citrate increased with a longer incubation time. Specific anion effects are detectable with neutral extractions.
It can be concluded, that the qualitative composition of root deposits has a higher influence on their P-mobilizing efficiency than their total amount.

Einleitung

Aus zahlreichen Arbeiten ist bekannt, dass die Rhizodeposition einen erheblichen Anteil am Energie- und Kohlenstoffhaushalt höherer Pflanzen ausmacht (MERBACH *et al.* 1999, KUZYAKOV and DOMANSKI 2000). Vor allem im wasserlöslichen Teil konnten verschiedene Verbindungen, in erster Linie Zucker und organische Säuren, identifiziert werden. Unter Stressbedingungen, beispielsweise unter P-Mangel, wurden sowohl quantitative als auch qualitative Verschiebungen im Spektrum der Wurzelabscheidungen deutlich (GRANSEE and WITTENMAYER 2000). Wasserlösliche organische Verbindungen unterliegen unter natürlichen Bodenbedingungen einer raschen mikrobiellen Umsetzung. Teilweise werden sie zu CO_2 abgebaut, der im Boden verbleibende Teil kann aber eine höhere Effektivität zur Phosphatmobilisierung aufweisen (MERBACH *et al.* 2000). Letztlich wurden auch bei *in vitro*-Kultur von Rhizosphärenbakterien potentiell P-mobilisierende Verbindungen identifiziert (DEUBEL *et al.* 2000), wobei neben verschiedenen niedermolekularen Carbonsäuren auch Zuckersäuren und Zuckeralkohole auffielen.

In der vorliegenden Arbeit soll daher geklärt werden, welche phosphatmobilisierende Wirkung solche Verbindungen in unterschiedlichen Böden haben. Damit können Hinweise gegeben werden, inwieweit Kenntnisse der Zusammensetzung der Rhizodeposition verschiedener Pflanzenarten Möglichkeiten zur Optimierung von Extraktionsverfahren zur P-Düngebedarfsermittlung bieten.

Material und Methoden

Das eingesetzte Extraktionsverfahren orientierte sich an der Doppellactat (DL)-Methode (HOFFMANN 1991). 2 g Boden wurden mit 0,2 g Aktivkohle und 100 ml Extraktionslösung geschüttelt. Es wurden Substanzkonzentrationen von einheitlich 5 g l^{-1} eingesetzt. Das bedeutete allerdings unterschiedliche molare Konzentrationen, bei organischen Säuren in Abhängigkeit von der Wertigkeit auch verschiedene Pufferkapazitäten der Extraktionslösungen. Tab. 1 gibt einen Überblick über die getesteten Substanzen sowie die für die Interpretation der Ergebnisse wichtigen molaren Konzentrationen.

Der pH-Wert der Lösungen wurde analog zur DL-Methode auf 3,7 eingestellt. Um einzuschätzen, inwieweit Mobilisierungseffekte durch spezifische Anionenwirkungen zu erklären sind, erfolgte parallel eine Extraktion mit neutralen Extraktionslösungen.

Die Extraktionszeit betrug wie bei den Standardmethoden 90 min. Um längerfristige, auch mikrobielle Umsetzungsprozesse zu erlauben, wurde zusätzlich eine 24 h – Extraktion bei 28 °C durchgeführt.

Tab. 1. Getestete Substanzen und Vorkommen solcher Verbindungen.

	mmol (meq) pro g Boden	
Glucose (Hauptbestandteil wasserlöslicher Wurzelabscheidungen (WA) krautiger Pflanzen)	1,4	
Galactose (WA verschiedener Pflanzen, steigender Anteil unter P-Mangel)	1,4	
Ribose (WA verschiedener Pflanzen, neben anderen Pentosen steigender Anteil unter P-Mangel)	1,7	
Xylitol (verschiedene Zuckeralkohole wurden bei in vitro-Kultur von Rhizosphären-bakterien identifiziert)	1,6	
Gluconsäure (mikrobieller Glucoseabbau)	1,3	(1,3)
Bernsteinsäure (WA vieler Pflanzen, mengenmäßig wichtiges bakterielles Stoffwechselprodukt)	2,1	(4,2)
Citronensäure (WA vor allem P-effektiver Pflanzen, wird von Bakterien und Pilzen produziert)	1,3	(3,9)
Oxalsäure (WA einiger Pflanzen, vor allem Beta-Arten)	2,8	(5,6)

Zum Vergleich wurden folgende Standardextraktionen herangezogen:
Doppellactat (**DL**): (6 g l^{-1} $C_6H_{10}O_6Ca \cdot 5\,H_2O$, 5 g Boden in 250 ml)
= 2,0 mmol Lactat / g Boden
Calciumacetat-Lactat (**CAL**): (15,4 g l^{-1} $C_6H_{10}O_6Ca \cdot 5\,H_2O$ + 7,5 g l^{-1} $(CH_3COO)_2Ca \cdot H_2O$ + 19 g l^{-1} CH_3COOH, 5 g Boden in 100 ml
= 2,0 mmol Lactat + 8,0 mmol Acetat / g Boden
H_2O (5 g Boden in 250 ml)

Die Extraktionen wurden an 10 verschiedenen Böden der Bodengruppen 3 und 5 (sandiger Lehm bzw. Ton) durchgeführt. Die pH-Werte lagen zwischen 5,2 (pH-Stufe C, VDLUFA Sachsen-Anhalt) und 7,6 (F). Die DL-löslichen P-Gehalte variierten zwischen 5,8 mg kg^{-1} Boden (Gehaltsklasse A) und 103,4 mg kg^{-1} (D). In Lösung befindliches Phosphat wurde im Filtrat nach MURPHY and RILEY (1962) gemessen.

Ergebnisse und Diskussion

Um eine gewisse Verallgemeinerbarkeit der Ergebnisse zu erreichen, wurden Mittelwerte über alle Böden, über Böden der Bodengruppe 3 bzw. 5 sowie über verschiedene Boden-pH-Bereiche gerechnet (Tab. 2).

Tab. 2. Mit unterschiedlichen Substanzen bei verschiedenen Extraktionszeiten und Lösungs-pH-Werten extrahierbare P-Mengen (mg P (kg Boden)$^{-1}$).

	Ø	**BG 3**	**BG 5**	**pH < 6**	**pH 6-7**	**pH > 7**
DL	52,7	54,1	49,7	64,9	52,3	29,3
CAL	29,1	28,6	30,1	33,2	28,0	23,0
H_2O	10,5	11,5	8,0	14,3	7,6	8,6
Glucose, pH 3,7 1,5h	23,3	23,9	21,9	31,5	21,7	10,2
Glucose, pH 3,7 24h	24,7	26,7	20,1	36,9	21,6	6,8
Glucose, pH 7,0 1,5h	23,7	24,4	22,2	33,0	21,7	9,2
Glucose, pH 7,0 24h	22,5	25,5	15,6	34,8	18,2	6,6
Ribose, pH 3,7 1,5h	26,8	27,2	25,9	35,5	24,8	13,5
Ribose, pH 3,7 24h	28,6	31,0	23,1	39,0	23,9	17,3
Ribose, pH 7,0 1,5h	26,0	26,2	25,4	33,6	25,6	11,6
Ribose, pH 7,0 24h	24,8	28,1	17,3	39,2	19,4	7,0
Gluconsäure, pH 3,7 1,5h	50,8	50,2	52,2	58,4	53,5	30,3
Gluconsäure, pH 3,7 24h	64,3	68,5	54,7	84,8	60,4	31,3
Gluconsäure, pH 7,0 1,5h	20,5	22,1	16,6	28,7	18,4	8,1
Gluconsäure, pH 7,0 24h	29,5	33,8	19,5	46,6	24,9	4,3
Bernsteinsäure, pH 3,7 1,5h	53,3	52,5	55,1	56,1	53,6	47,1
Bernsteinsäure, pH 3,7 24h	56,1	61,5	43,4	74,2	45,3	41,4
Bernsteinsäure, pH 7,0 1,5h	11,8	12,3	10,7	15,8	11,0	5,5
Bernsteinsäure, pH 7,0 24h	16,3	18,1	12,1	26,1	13,4	2,5
Citronensäure, pH 3,7 1,5h	163,2	165,3	158,2	185,3	164,1	117,0
Citronensäure, pH 3,7 24h	316,6	314,8	320,7	351,0	336,4	208,0
Citronensäure, pH 7,0 1,5h	72,2	74,3	67,2	87,8	69,9	45,5
Citronensäure, pH 7,0 24h	187,2	198,7	160,3	235,6	185,3	94,3
Oxalsäure, pH 3,7 1,5h	244,3	242,4	248,8	275,8	254,9	160,3
Oxalsäure, pH 3,7 24h	327,7	300,1	392,0	340,1	364,9	228,5
Oxalsäure, pH 7,0 1,5h	101,4	110,3	80,7	130,0	81,5	84,0
Oxalsäure, pH 7,0 24h	134,0	143,3	112,2	164,1	120,9	99,8

Glucose mobilisierte durchschnittlich 40-45 % der mit DL extrahierbaren P-Menge. Da Glucoselösung keine Pufferkapazität aufweist, spielte der eingestellte pH-Wert keine Rolle. In sauren Böden wurde mit 50-55 % von DL die Größenordnung der CAL-Extraktion erreicht, auf alkalischen Böden mit 20-25 % dagegen nur die der Wasserextraktion. Galactose und Xylitol ergaben ein nahezu dekkungsgleiches Bild mit Glucose, daher wurde in Tab. 2 auf eine Darstellung der Ergebnisse verzichtet. Ribose war mit durchschnittlich 50 % von DL, bis zu 60 % auf sauren Böden, etwas effektiver, zeigte aber ebenfalls wenig Wirkung auf alkalischen Böden.

Die Wirkung von Gluconsäure war bei saurer Extraktion (pH 3,7) trotz niedrigerer Pufferkapazität vergleichbar mit DL. Auf sauren und leichteren Böden war ein deutlicher Anstieg bei längerer (24 h) Extraktion zu beobachten. Bei neutraler Extraktion ähnelte die Gluconatwirkung der der Zucker, d.h. auch Gluconat hatte keine Wirkung auf alkalischen Böden. Eine chelatisierende Wirkung (WHITELAW *et al.* 1999) scheint, wenn überhaupt, nur unter sauren Bedingungen möglich.

Bernsteinsäureextraktion entsprach bei pH 3,7 im Durchschnitt ebenfalls der DL-Methode; bei alkalischen Böden führte die höhere Pufferkapazität zu besseren Ergebnissen. Eine neutrale Succinatlösung war dagegen nicht effektiver als Wasser. Damit dürfte die Bedeutung der Bernsteinsäure in der Rhizosphäre auf eine Absenkung des pH-Wertes beschränkt sein, spezifische Anioneneffekte wurden nicht beobachtet.

Citronen- und Oxalsäure unterschieden sich extrem von allen anderen untersuchten Substanzen. Bei gleichen Extraktionsbedingungen setzte Citronensäure ca. 300 % der DL-P-Mengen frei. Eine 24 h – Extraktion brachte eine Verdoppelung auf 600 %, wobei in diesem Ansatz nicht geklärt werden kann, inwieweit dafür rein chemische oder aber mikrobielle Umsetzungsprozesse verantwortlich sind. Eine neutrale Citratextraktion setzte zwar wesentlich weniger P frei, war aber der sauren DL-Extraktion immer noch deutlich überlegen. Noch stärker fiel hier der Anstieg in 24 h aus. Oxalsäure zeigte bei gleicher Massekonzentration in der Extraktionslösung die höchste Effizienz (durchschnittlich 460 % von DL in 90 min, 600 % von DL in 24 h). Besonders wirksam war Oxalsäure bei pH 3,7 auf schweren Böden. Eine neutrale Extraktion brachte immerhin 200-250 % der DL-Extraktion.

Um festzustellen, welchen Einfluss die Substanzkonzentration auf die P-Freisetzungsrate dieser hocheffektiven Verbindungen hat, wurden Extraktionen mit Citronen- bzw. Oxalsäurekonzentrationen zwischen 0 und 5 g l^{-1} durchgeführt (pH 3,7, 90 min). Dafür wurden 4 verschiedene Böden (alle BG 3), zwei saure und zwei alkalische, jeweils mit hohem bzw. niedrigem DL-P-Gehalt, eingesetzt. Die Ergebnisse sind in Abb. 1 und 2 dargestellt.

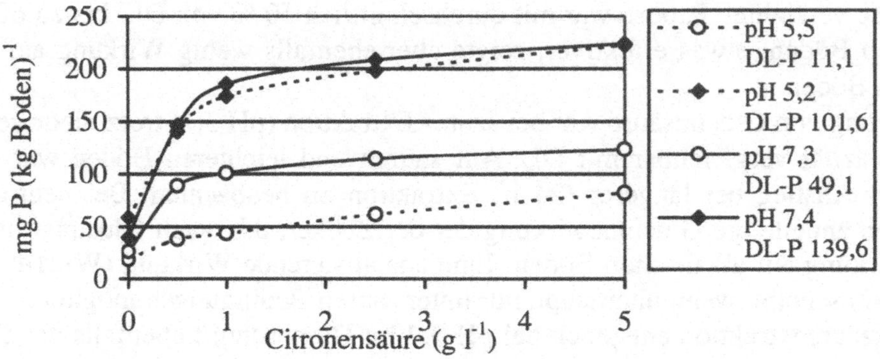

Abb. 1. Einfluss der Citronensäurekonzentration auf die P-Freisetzung aus unterschiedlichen Böden (zum Vergleich jeweils DL-löslicher P angegeben).

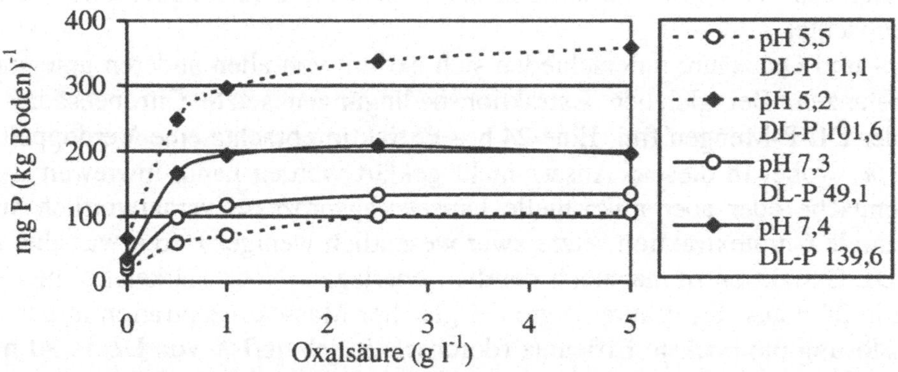

Abb. 2. Einfluss der Oxalsäurekonzentration auf die P-Freisetzung aus unterschiedlichen Böden (zum Vergleich jeweils DL-löslicher P angegeben).

Die Abbildungen 1 und 2 zeigen, dass die P-Freisetzung mit steigender Citrat- oder Oxalatkonzentration im Extraktionsmittel keineswegs linear steigt, sondern dass sich deutliche Sättigungskurven ergeben. Von 0 auf 0,5 g l^{-1} war ein steiler Anstieg der P-Freisetzungsrate zu verzeichnen. Bereits in dieser Konzentration (ein Zehntel der ursprünglich eingesetzten Mengen!), die 0,13 mmol Citrat bzw. 0,28 mmol Oxalat pro g Boden entspricht, wurde in jedem Fall mehr Phosphat mobilisiert als mit der DL-Extraktion. Oberhalb einer Konzentration von 1 g l^{-1} war der Kurvenanstieg dagegen gering. Das zeigt einerseits, dass Citrat und Oxalat bereits in geringen Konzentrationen, die in der Rhizosphäre effizienter Pflanzen durchaus erreicht werden (JONES 1998), die P-Verfügbarkeit erheblich verbes-

sern. Andererseits hätte auch ein Vergleich der Substanzen auf gleicher molarer (7,7 g l^{-1} Citronensäure, 3,6 g l^{-1} Oxalsäure) oder normaler Konzentrationsebene (2,6 bzw. 1,8 g l^{-1}) nicht zu wesentlich anderen Ergebnissen geführt.
Während die P-Freisetzungskurven mit Citrat und Oxalat in den beiden alkalischen Böden ähnlich verliefen, war Oxalat, zumindest mit der hier gewählten 90minütigen Extraktionszeit, auf sauren Böden wesentlich effizienter. Oxalat scheint besonders sorbierte Phosphate effektiv zu mobilisieren, was sich auch in der guten Wirkung auf schweren, sorptionsstarken Böden zeigt.

Fazit

- Zucker und Zuckeralkohole können unter unsterilen Bedingungen bei Boden-pH-Werten unter 7 nennenswert zur P-Mobilisierung beitragen, zeigten aber keine Effekte unter alkalischen Bodenbedingungen.
- Gluconsäure war trotz geringerer Pufferkapazität bei pH 3,7 vergleichbar mit DL, bei einem deutlichen Anstieg in 24 h. Bei neutraler Extraktion ähnelte die Wirkung der der Zucker.
- Bernsteinsäure war bei pH 3,7 vergleichbar mit DL, bei pH 7 nur mit Wasserextraktion, Effekte wurden hauptsächlich durch pH-Absenkung erreicht.
- Citrat und Oxalat könnten schon in geringen Konzentrationen aus unterschiedlichsten Böden (Oxalat vor allem bei sauren Böden) erhebliche P-Mengen freisetzen. Vor allem die Effektivität von Citrat stieg deutlich bei längerer Inkubationszeit. Spezifische Anioneneffekte wirken auch bei neutraler Extraktion.

Die qualitative Zusammensetzung in der Rhizosphäre vorhandener organischer Verbindungen beeinflusst die P-mobilisierende Wirkung der Rhizodeposition stärker als deren Menge.
Die P-Aufnahme von Pflanzen, die hocheffektive Verbindungen, wie Citrat und Oxalat, produzieren, lässt sich mit Standardmethoden kaum abschätzen. Eine Extraktion mit solch hochwirksamen Verbindungen könnte in Ergänzung zu gängigen Standardmethoden das Potenzial mobilisierbarer P-Verbindungen in Böden bestimmen, reflektiert aber nicht die pflanzenverfügbare P-Menge für weniger P-effiziente Pflanzenarten.

Literaturverzeichnis

DEUBEL, A.; GRANSEE, A.; MERBACH, W., 2000: Transformation of organic rhizodepositions by rhizosphere bacteria and its influence on the availability of tertiary calcium phosphate. *Journal of Plant Nutrition and Soil Science* 163, 387-392.

GRANSEE, A.; WITTENMAYER, L., 2000: Qualitative and quantitative analysis of water-soluble root exudates in relation to plant species and development. *Journal of Plant Nutrtition and Soil Science* 163, 381-385.

HOFFMANN, G., 1991: „1. Die Untersuchung von Böden," VDLUFA-Verlag, Darmstadt.

JONES, D. L., 1998: Organic acids in the rhizosphere - a critical review. *Plant and Soil* 205, 25-44.

KUZYAKOV, Y.; DOMANSKI, G., 2000: Carbon input by plants into the soil. Review. *Journal of Plant Nutrition and Soil Science* 163, 421-431.

MERBACH, W.; GRANSEE, A.; DEUBEL, A., 2000: Phosphate mobilisation by root exudates and rhizosphere microbes. *Plant Physiology and Biochemistry* 38, 13-21.

MERBACH, W.; MIRUS, E.; KNOF, G.; REMUS, R.; RUPPEL, S.; RUSSOW, R.; GRANSEE, A.; SCHULZE, J., 1999: Release of carbon and nitrogen compounds by plant roots and their possible ecological importance. *Journal of Plant Nutrition and Soil Science* 162, 373-383.

MURPHY, J.; RILEY, J. P., 1962: A modified single solution method for the determination of phosphate in natural waters. *Analytica Chimica Acta* 27, 31-36.

WHITELAW, M. A.; HARDEN, T. J.; HELYAR, K. R., 1999: Phosphate solubilization in solution culture by the soil fungus *Penicillium radicum*. *Soil Biology and Biochemistry* 31, 655-665.

5

Kohlenstoff- und Stickstoffumsatz im durchwurzelten Bodenraum

Prozessregulation in der Rhizosphäre
13. Borkheider Seminar zur Ökophysiologie des Wurzelraumes
Hrsg.: W. Merbach, B. W. Hütsch, L. Wittenmayer, J. Augustin
B. G. Teubner Stuttgart, Leipzig, Wiesbaden (2003), S. 125-128

Einfluss des Pflanzenwachstums und der organischen Substanz auf C- und N- Umsatz im Boden

Abdel R.A. USMAN, Yakov KUZYAKOV und Karl STAHR
Institut für Bodenkunde und Standortslehre (310), Universität Hohenheim, D-70593 Stuttgart

Abstract

An artificial soil was produced by open air composting of a mixture of sewage sludge, wood shavings, green compost, and a sandy loam soil, and used to investigate the effect of plant growth (wheat) and organic matter application on turnover of organic C und N. A greenhouse experiment was carried out with and without plants (wheat) to test the turnover of N in artificial soil and a mixture of artificial soil with sand in the ratio of 1:1. The results showed that the cultivation of wheat increased the net N mineralization by about 68% to 71% in comparison to treatments without plants. A laboratory experiment was carried out to investigate the effect of incubation of straw or animal manure with artificial soil at 35 °C on C mineralization and net N mineralization. The addition of straw and animal manure caused an increase in mineralization of C. The net N mineralization was higher after addition of animal manure and without substrate application (control) than after addition of straw.

Einleitung

Im Rahmen dieser Arbeit wurde ein künstlicher Boden, der durch Inkubation von Klärschlamm, Pflanzenresten, Holz und sandigem Lehm hergestellt wurde, für Untersuchungen zum C- und N-Umsatz herangezogen. Einer der Faktoren, der den N- und C-Umsatz in der organischen Bodensubstanz beeinflusst, ist der Anbau von Pflanzen. Pflanzenanbau kann entweder zur Beschleunigung oder zur Verlangsamung des Abbaus der organischen Bodensubstanz führen (KUZYAKOV 2002). Das Ziel dieser Arbeit ist einerseits die Untersuchung der Auswirkung des Anbaus von Weizen auf den N-Umsatz in der organischen Bodensubstanz des künstlichen Bodens und andererseits die Prüfung der Effekte der Inkubation von Stroh und Stallmist mit künstlichem Boden auf den C- und N-Umsatz.

Material und Methoden

Der N-Umsatz der organischen Bodensubstanz des künstlichen Bodens und eines Gemisches aus künstlichem Boden mit Quarzsand im Verhältnis 1:1 wurde im Gefäßversuch mit und ohne Weizen geprüft. Der künstliche Boden hatte folgende Charakteristika: Corg (4,84%), N_t (0.23%), C/N-Verhältnis (21.5) und N_{min} (0.19 g kg^{-1}). Die N-Nettomineralisation wurde durch Differenzbildung zwischen der Summe aus N_{min} und N-Aufnahme durch Weizen nach einer Anbauzeit von 50 Tagen und N_{min} vor dem Anbau, mit und ohne Pflanzen, berechnet.
Im einem Laborversuch wurde die Inkubation von organischer Substanz (Stroh oder Stallmist) mit künstlichem Boden bei 35 °C durchgeführt, um die Auswirkung des Strohs und des Stallmistes auf den C- und N-Umsatz zu untersuchen. Es wurden jeweils 40 g Boden mit Weizenstroh und Stallmist (250 mg C 100 g^{-1} Boden) gemischt. Die Inkubationszeit betrug 110 Tage. Die CO_2–Gesamteffluxe und die N-Nettomineralisation des künstlichen Bodens wurden mit und ohne Zugabe von Stroh und Stallmist gemessen.

Ergebnisse und Diskussion

Gefäßversuch:

N-Nettomineralisation

Mit Weizen steigt die N-Nettomineralisation im künstlichen Boden und im künstlichen Boden+Sand Gemisch im Vergleich zu Boden ohne Bewuchs an (Abb. 1). Im Vergleich zu Boden ohne Pflanzen führt der Anbau von Weizen zu einer zusätzlichen N-Nettomineralisation von 13.7 kg ha^{-1} Tag^{-1} bei künstlichem Boden und 3.2 kg $ha^{-1}Tag^{-1}$ bei künstlichem Boden+Sand. Nach 50tägigem Weizenanbau betrugen die Prozentgehalte des Netto-N_{min} am Gesamt-N-Gehalt bei künstlichem Boden mit Pflanzen 14.4%, bei künstlichem Boden ohne Pflanzen 4.5%, bei künstlichem Boden+Sand mit Pflanzen 3.3%, und bei künstlichem Boden+Sand ohne Pflanzen 0.98%. Das Gemisch von Quarzsand mit künstlichem Boden verringerte die N-Nachlieferung. Die Erklärung hierfür ist in der Verringerung des Gehalts an N_t und in der Veränderung der Eigenschaften des künstlichen Bodens durch Verdünnung mit Quarzsand zu sehen. Der Anbau von Weizen förderte

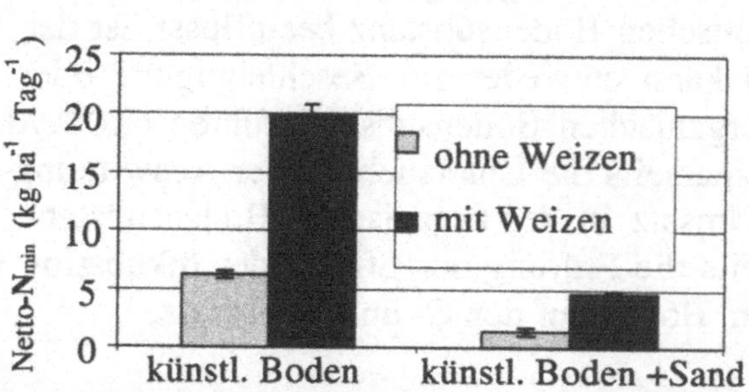

Abb. 1. N-Nettomineralisation im künstlichen Boden und mit 50% künstlichem Boden +50% Sand, mit und ohne Weizen.

wahrscheinlich die mikrobielle Aktivität und damit erfolgte ein zusätzlicher Abbau der organischen Bodensubstanz.

Inkubationsversuch:
1. C-Mineralisierung

Im Vergleich zum Boden ohne Zugabe von Stroh und Stallmist wurde ein erheblicher Anstieg der CO_2-Gesamteffluxe aus dem künstlichen Boden nach Applikation von Stroh und Stallmist festgestellt (Abb. 2). Am Ende der Inkubationszeit (nach 110 Tage) führte die Zugabe von Stroh zur 1.7fachen und die Zugabe von Stallmist zur 1.5fachen C-Freisetzung im Vergleich zum Boden ohne Zugabe.

Die Mengen des freigesetzten C sind bei Stallmist (38.2 % von zugegebenem C_{org}) geringer als bei Stroh (53.7 % von zugegebenem C_{org}).

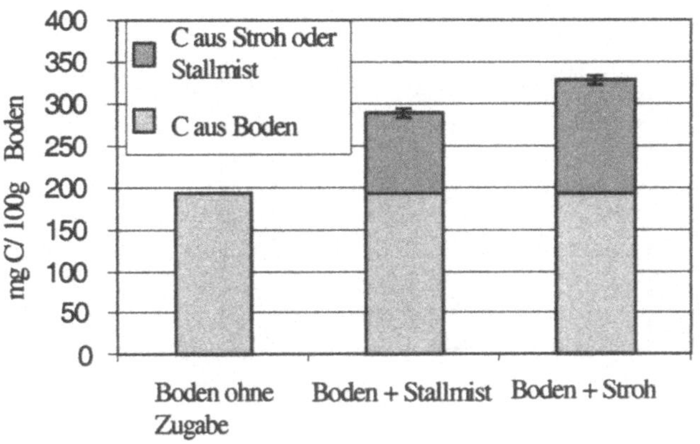

Abb. 2. CO_2-Gesamtefflux aus dem künstlichen Boden mit und ohne Zugabe von Stroh oder Stallmist nach 110 Tagen Inkubationszeit.

2. CO_2-Effluxraten

Die CO_2–Effluxraten waren maximal zu Beginn des Versuches (innerhalb der ersten 20 Tage) (Abb. 3). Die Zugabe von Stroh bewirkte höhere CO_2-Flüsse als die Zugabe von Stallmist (besonders innerhalb der ersten 2 Wochen).

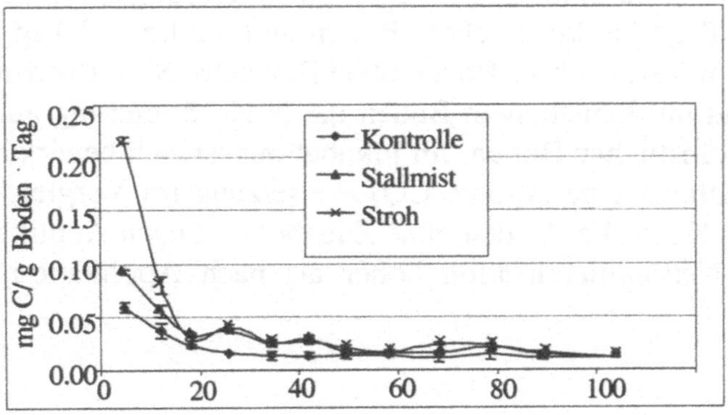

Abb. 3. CO_2–Effluxraten im Verlauf des Inkubationsversuches (110 Tage) mit und ohne Zugabe von Stroh oder Stallmist.

3. N-Nettomineralisation

Die Kontrolle (Boden ohne Zugabe) und der Boden mit Zugabe von Stallmist verursachten eine höhere N-Nettomineralisation nach 110tägiger Inkubation im Vergleich zu Boden mit Zugabe von Stroh (Abb. 4). Die geringste N-Nettomineralisation durch Zugabe von Stroh ist damit zu begründen, dass für den mikrobiellen Abbau des Strohs wegen dessen weitem C/N-Verhältnis zusätzlicher Stickstoff benötigt wird. Am Ende der Inkubation (nach 110 Tagen) ergaben die Prozentgehalte an Netto-N_{min} vom Gesamt-N-Gehalt für die Kontrolle (Boden ohne Zugabe) 44.5%, für Boden mit Zugabe von Stroh 25.1% und für Boden mit Zugabe von Stallmist 46.9%.

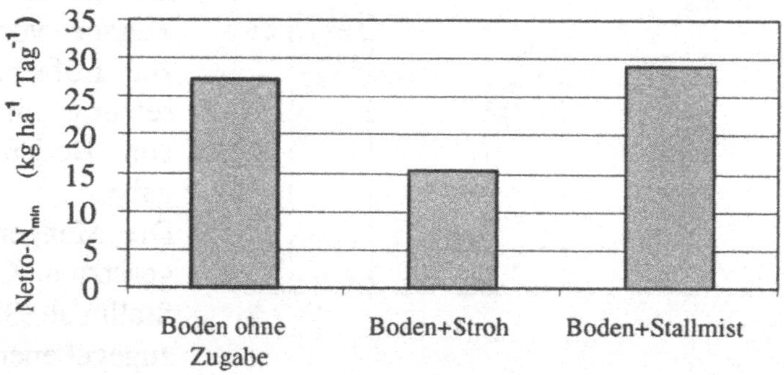

Abb. 4. N-Nettomineralisation nach 110tägiger Inkubation des Bodens ohne Zugabe oder nach Applikation von Stroh oder Stallmist.

Schlussfolgerungen

Der Anbau von Weizen führte zu einer zusätzlichen Nachlieferung an mineralischem N von 13.7 kg ha^{-1} Tag^{-1} bei künstlichem Boden und 3.2 kg ha^{-1} Tag^{-1} bei künstlichem Boden+Sand im Vergleich zu Boden ohne Bewuchs. Somit verringert ein Gemisch aus Quarzsand mit künstlichem Boden die N-Nachlieferung auf ein Viertel im Vergleich zum künstlichen Boden. Im Inkubationsversuch bewirkte die Zugabe von Stroh und Stallmist eine erhöhte CO_2-Freisetzung im Vergleich zu Boden ohne Zugabe. In der Kontrolle (Boden ohne Zugabe) und nach Applikation von Stallmist war die N-Nettomineralisation höher als nach Applikation von Stroh.

Literaturverzeichnis

KUZYAKOV, Y., 2002: Review: Factors affecting rhizosphere priming effects. Journal of Plant Nutrition and Soil Science 165, 382-396.

Prozessregulation in der Rhizosphäre
13. Borkheider Seminar zur Ökophysiologie des Wurzelraumes
Hrsg.: W. Merbach, B. W. Hütsch, L. Wittenmayer, J. Augustin
B. G. Teubner Stuttgart, Leipzig, Wiesbaden (2003), S. 129-135

Photosynthesis as a factor controlling short-term C turnover in the rhizosphere

Yakov KUZYAKOV[1] und Weixin CHENG[2]
[1] Institute of Soil Science and Land Evaluation, University of Hohenheim, Emil-Wolff-Straße 27; D-70599 Stuttgart, Germany
[2] Environmental Studies, University of California - Santa Cruz, Santa Cruz, CA 95064, USA

Abstract

The effect of different shading periods of wheat or maize on short-term changes of soil organic matter (SOM) decomposition in the rhizosphere was investigated by using ^{13}C natural abundance. Cultivation of wheat (C_3 plant) on Haplic Chernozem (C_4 soil) increased SOM decomposition intensity (positive priming effect). ^{13}C natural abundance allows tracing of the dynamics of the priming effect depending on the light-on events.
In contrast to wheat, cultivation of maize (C_4 plant) on Chromic Luvisol (C_3 soil) decreased SOM mineralization compared to unplanted soil (negative priming effect). Prolonged and strongly prolonged dark periods reduced the changes of SOM decomposition in the rhizosphere induced by maize under normal day/night cycles. Both investigated plant-soil pairs show tight coupling of short-term SOM decomposition with photosynthesis intensity.

Introduction

Some investigations of CO_2 efflux from soil under natural conditions have shown diurnal patterns in the efflux rates (BALDOCCHI *et al.* 1986, KIM and VERMA 1992). Most investigators have attributed these diurnal fluctuations to diurnal soil temperature changes because soil temperature has repeatedly been shown to be one of the important controlling factors for soil CO_2 efflux (FANG and MONCRIEFF 2001). Plant photosynthesis has rarely been considered as an important controlling factor for the diurnal fluctuation of soil CO_2 efflux, even though substrate supply for rhizosphere processes is controlled by photosynthesis. High transport rates of assimilates from leaves into the roots and then the loss by root respiration and exudation of organic substances into the rhizosphere have been reported based on data from laboratory experiments (GREGORY and ATWELL 1991, CHENG *et al.* 1993). As shown in previous investigations, root-derived CO_2 efflux as a sum of microbial respiration of exudates (rhizomicrobial respiration) and root respiration strongly depends on day/night period (KUZYAKOV and CHENG 2001). In the later study separating root and rhizomicrobial respiration, KUZYAKOV and SINIAKINA (2001) showed that diurnal photosynthesis cycles affected exudation much stronger than root respiration. Therefore it can be expected, that changes of microbial activity in the rhizosphere by exudation induced by photosynthesis will result in modification of microbial decomposition of soil organic matter (SOM). Therefore, the aim of this study was to investigate the effect of photosynthesis on SOM decomposition.

Material and Methods

In order to follow this aim, two experiments were conducted including growth of a C_3 plant (spring wheat) on a C_4 soil (Haplic Chernozem from tallgrass Kansas Prairie, $\delta^{13}C = -14.85‰$) or *vice versa* (maize on a Chromic Luvisol $\delta^{13}C = -24.45‰$ from California sea terrace with grassland) and monitoring $\delta^{13}C$ values in CO_2 efflux from planted and unplanted soil.

Five germinated seedlings for wheat and one seed for maize were transplanted in PVC containers with 1 kg soil. The plants were grown in a growth chamber at a constant (22 ± 0.5 °C) day and night temperature with a light intensity of 800 μmol m^{-2} s^{-1} and a soil water content of 80% of the available field capacity. Before the start of light treatments (d 31 for wheat, d 14 for maize) all plants were grown under 12/12 h day/night periods.

Two day/night settings were investigated in the experiment with wheat: The 'normal' day/night period had a day-length of 12 h and a night-length of 12 h. The 'prolonged' night had a day-length of 12 h and a prolonged night of 60 h. Two of these prolonged night cycles (2 × 72 h), were investigated with wheat. Three day/night settings were investigated in the experiment with maize: The 'normal' day/night period had a day/night-length of 12/12 h. The 'prolonged' night had a day-length of 12 h and a prolonged night of 36 h. The 'strongly prolonged' night had a day-length of 12 h and a prolonged night of 84 h. Eight normal cycles (8 × 24 h), four prolonged (4 × 48 h) and 2 strongly prolonged (2 × 96 h) day/night periods were investigated with maize.

$\delta^{13}C$ values in CO_2 evolution from soil with or without plants as well as in root tissue and soil were measured. The total CO_2 efflux from the root-soil columns was partitioned into the root-derived and the SOM-derived parts using ^{13}C natural abundance method (CHENG 1996):

$$C_{Root} = C_t \times (\delta_t - \delta_{SOM}) / (\delta_{Root} - \delta_{SOM})$$

where $C_t = C_{SOM} + C_{Root}$, is the total C from below ground CO_2, C_{SOM} is the amount of C-CO_2 derived from the SOM, C_{Root} is the amount of C-CO_2 derived from roots, δ_t, δ_{Root} and δ_{SOM} are the $\delta^{13}C$ values of the C_t, root C, and CO_2 evolved from the soil without plants (=SOM derived CO_2). The SOM-derived CO_2 efflux from planted soil was calculated as the difference between total CO_2 efflux (C_t) and root-derived CO_2 (C_{Root}) obtained by the ^{13}C natural abundance method. The change of SOM-decomposition due to plant roots was calculated as the difference between SOM-derived CO_2 from planted soil and from unplanted soil.

The data are presented as means of four replicates ± standard deviation (SD). t-Test ($\alpha \leq 5\%$) was used to indicate the significance of differences between treatments. Both experiments are described in detail by KUZYAKOV and CHENG (2001, 2003).

Results and Discussion

The use of the ^{13}C natural abundance method (CHENG 1996) permitted the separation of the SOM-derived and root-derived CO_2 from planted soil. SOM-derived CO_2 efflux varied between 5% and 50% of total CO_2 efflux from the soil rooted with wheat (Fig. 1). On average the SOM-derived CO_2 efflux from the planted soil amounted to 25% of the total CO_2 efflux. Two light-on events d 1 and d 4 led to strong increases of total CO_2 efflux from the soil. These increases lasted about

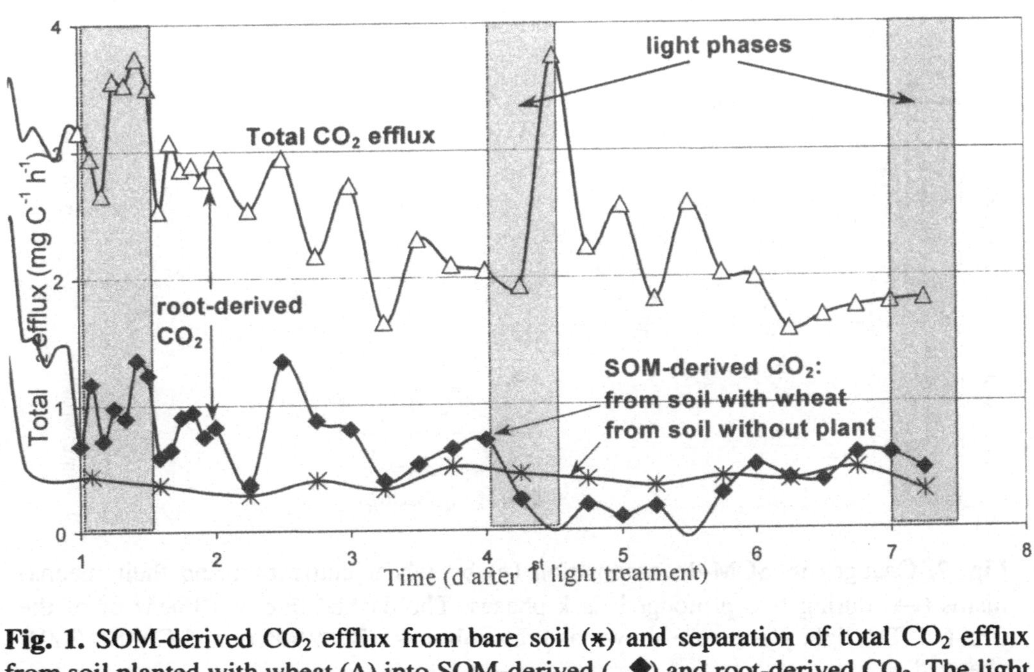

Fig. 1. SOM-derived CO_2 efflux from bare soil (∗) and separation of total CO_2 efflux from soil planted with wheat (Δ) into SOM-derived (◆) and root-derived CO_2. The light phases are shown as gray columns (prolonged dark period) (KUZYAKOV and CHENG 2001, changed).

12 h. Then the CO_2 efflux intensity decreased to the previous rate. The increase of CO_2 efflux after the second light period was greater than after the first period. The decrease after the second light period was faster than after the first one (Fig. 1). During the first half of the observation period the SOM-derived CO_2 from the planted soil was higher than from the unplanted soil.

At the beginning of the second half the SOM-derived CO_2 from the planted soil was less than from the unplanted soil (Fig. 1). The difference between the SOM-derived CO_2 from the planted and unplanted soil was a measure of the additional SOM mineralization caused by root growth (Fig. 2). This phenomenon - short-term change of SOM decomposition intensity - is frequently called 'priming effect' (JENKINSON *et al.* 1985, KUZYAKOV *et al.* 2000, KUZYAKOV 2002). It occurs often in the rhizosphere. Normal intensity of photosynthesis (12/12 h) led to exudation of easily decomposable exudates and resulted in increased microbial growth and activity in the rhizosphere (biomass activation). Subsequently SOM decomposition increased and a higher amount of nutrients was mineralized. Therefore, the priming effect was positive during the first 3 d and amounted to approximately 0.42 mg C kg^{-1} h^{-1} (Fig. 2), or 33 kg C ha^{-1} d^{-1} (calculated for 30 cm

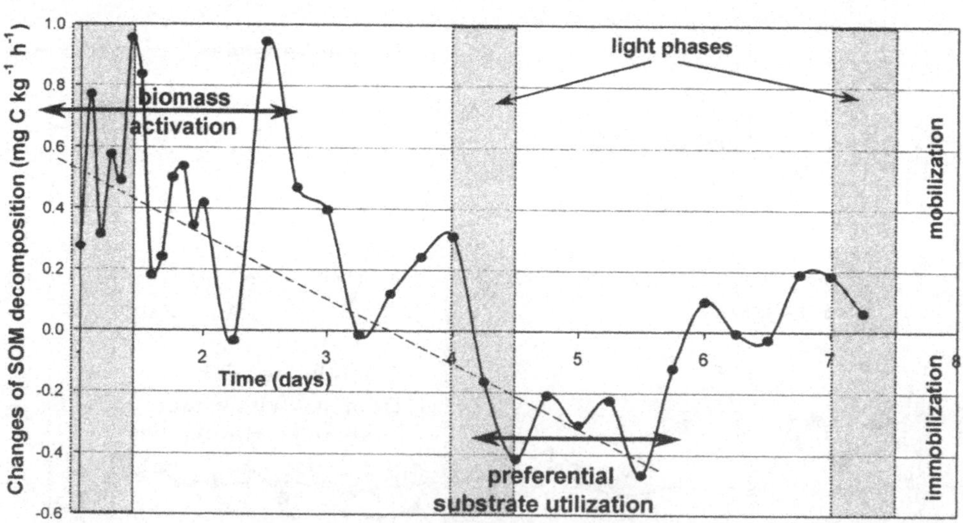

Fig. 2. Changes in SOM-decomposition (•) by wheat cultivation and their mechanisms (↔) during two prolonged dark phases. The dashed line is a linear fit of the first 6 d. The light phases are shown as gray columns. (KUZYAKOV and CHENG 2001, changed).

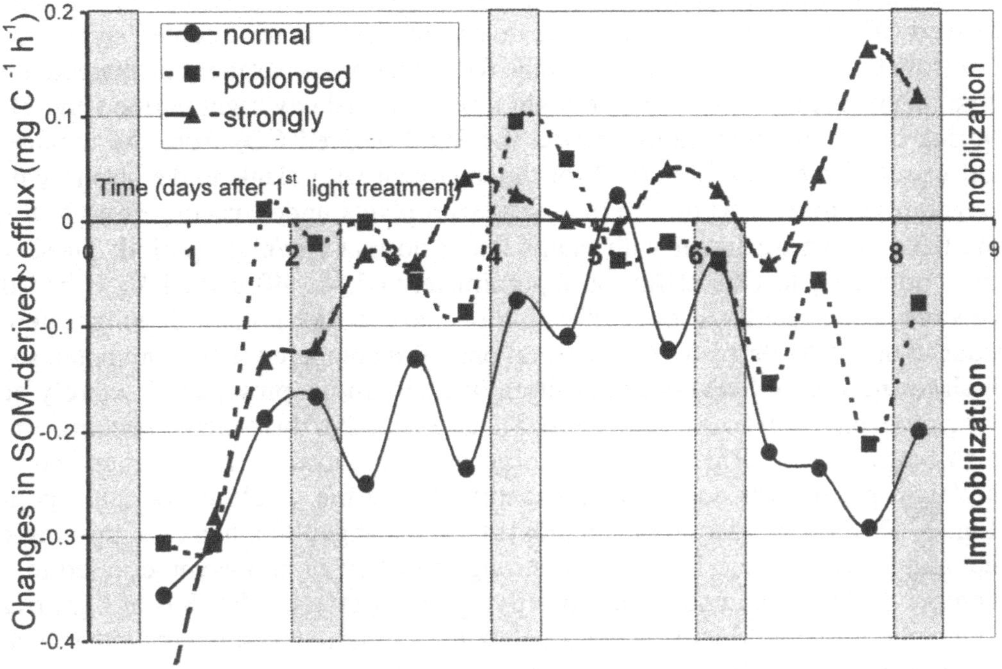

Fig. 3. Changes in SOM-decomposition induced by maize grown under normal (● : 12/12 h), prolonged (■ : 12/36 h), and strongly prolonged (▲ : 12/80 h) dark period. The light phases for prolonged and strongly prolonged darkness treatments are shown as gray columns. (KUZYAKOV and CHENG 2003).

soil layer and 1.1g cm^{-3} soil density). However, the priming effect decreased during 2.5d without light. The light-on event on d 4 led to the switch of microorganisms which previously utilized SOM on the easily available new exudates (pre-ference substrate utilization) and resulted in retardation of SOM decomposition.

In contrast to wheat, maize (C_4 plant) cultivation led to a decrease of SOM decomposition under normal day/night period (Fig. 3). Prolongation of dark period increased SOM decomposition. During the first day of measurement, SOM-derived CO_2 from the planted soil was much lower than from the unplanted soil (Fig. 3). One day without light increased the SOM-derived CO_2 and two days without light led to similar rates of SOM-derived CO_2 from planted and unplanted soil. SOM-derived CO_2 efflux in the experiment could be divided into three periods. During the first two days, the SOM-derived CO_2 from planted soil was much lower than that from the unplanted soil. Between d 3 and d 6 the SOM-derived CO_2 efflux from the soil with reduced photosynthesis was similar to that of the unplanted soil. After d 6 there were significant differences in the SOM-derived CO_2 from the soil of all four treatments and the treatments can be arranged from the highest to the lowest in the following order: strongly prolonged night time > unplanted soil > prolonged night time > normal day/night period (Fig. 3). During the whole observation period the SOM-derived CO_2 from the unplanted soil was 0.57 mg C kg^{-1} h^{-1}, 0.53 for the treatment with plants under strongly prolonged night time, 0.50 for the treatment with plants under prolonged night time, and 0.39 for the treatment with plants under normal day/night period. These values correspond to CO_2 efflux of approximately 45, 42, 40 and 31 kg C ha^{-1} d^{-1}. This result indicated that the cultivation of maize under normal day/night period reduced the SOM-derived CO_2 by about one third compared to the unplanted soil. Such retardation of SOM decomposition intensity in the presence of actively growing roots has been frequently referred to as a negative rhizosphere priming effect (KUZYAKOV 2002). Rhizosphere priming effect induced by maize under normal day/night period was consistently negative during the whole observation period (Fig. 3, 4). Strong reduction of photosynthesis changed this negative priming effect and it became zero later. 7 d of strong reduction in photosynthesis led to the increase d SOM decomposition – positive priming effect. After 8 d of light treatments, the priming effect was the greatest for strong reduced photosynthesis, and the smallest for the plants with normal day/ night period (Fig. 4).

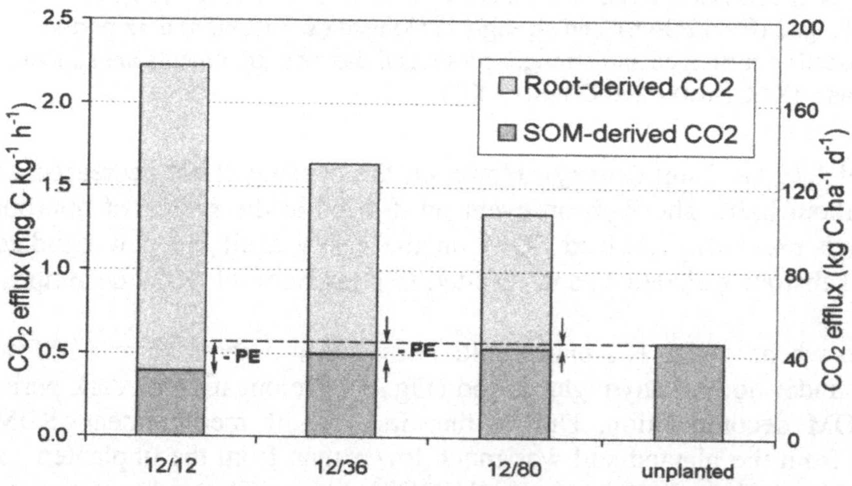

Fig. 4. Average intensity of CO_2 efflux from unplanted soil and soil planted with maize under three different light regimes and reduction of SOM-decomposition (**- PE**) induced by maize. (KUZYAKOV and CHENG 2003).

References

BALDOCCHI, D. D.; VERMA, S. B.; MATT, D. R.; ANDERSON, D. E., 1986: Eddy-correlation measurements of carbon dioxide efflux from the floor of a deciduous forest. *Journal of Applied Ecology* 23, 967-975.

CHENG, W., 1996: Measurement of rhizosphere respiration and organic matter decomposition using natural ^{13}C. *Plant and Soil* 183, 263-268.

CHENG, W.; COLEMAN, D. C.; CARROLL, C. R.; HOFFMAN, C. A., 1993: *In situ* measurement of root respiration and soluble C concentrations in the rhizosphere. *Soil Biology and Biochemistry* 25, 1189–1196.

FANG, C.; MONCRIEFF, J. B., 2001: The dependence of soil CO_2 efflux on temperature. *Soil Biology and Biochemistry* 33, 155-165.

GREGORY, P. J.; ATWELL, B. J., 1991: The fate of carbon in pulse labelled crops of barley and wheat. *Plant and Soil* 136, 205-213.

JENKINSON, D. S.; FOX, R. H.; RAYNER, J. H., 1985: Interactions between fertilizer nitrogen and soil nitrogen - the so-called 'priming' effect. *Journal of Soil Science* 36, 425-444.

KIM, J.; VERMA, S. B., 1992: Soil surface CO_2 flux in a Minnesota peatland. *Biogeochemistry* 18, 37-51.

KUZYAKOV, Y., 2002: Review: Factors affecting rhizosphere priming effects. *Journal of Plant Nutrition and Soil Science* 165, 382-396.

KUZYAKOV, Y.; CHENG, W., 2003: Photosynthesis controls of CO_2 efflux from maize rhizosphere. *Plant and Soil* (submitted).

KUZYAKOV, Y.; CHENG, W., 2001: Photosynthesis controls of rhizosphere respiration and organic matter decomposition. *Soil Biology and Biochemistry* 33, 1915-1925.

KUZYAKOV, Y.; SINIAKINA, S. V., 2001: Siphon method of separating root-derived organic compounds from root respiration in non-sterile soil. *Journal of Plant Nutrition and Soil Sciemce* 164, 511-517.

KUZYAKOV, Y.; FRIEDEL, J. K.; STAHR, K., 2000: Review of mechanisms and quantification of priming effects. *Soil Biology and Biochemistry* 32, 1485-1498.

Prozessregulation in der Rhizosphäre
13. Borkheider Seminar zur Ökophysiologie des Wurzelraumes
Hrsg.: W. Merbach, B. W. Hütsch, L. Wittenmayer, J. Augustin
B. G. Teubner Stuttgart, Leipzig, Wiesbaden (2003), S. 136-137

Untersuchung des Einflusses von Ureaseinhibitoren auf die pflanzliche Verwertung von harnstoffbürtigem Stickstoff

Volker HOFFMANN (†), Wolfgang GANS und Wolfgang MERBACH
Institut für Bodenkunde und Pflanzenernährung der Martin-Luther-Universität Halle-Wittenberg, Adam-Kuckhoff-Straße 17 b, D- 06108 Halle/Saale

Zusammenfassung

Harnstoff nimmt im Rahmen der Stickstoffdüngemittel im Weltmaßstab einen immer breiteren Raum ein. Wird Harnstoff als Dünger ausgebracht, kommt es relativ schnell zur Zersetzung zu Ammoniak, Wasser und Kohlendioxid. Verantwortlich dafür ist das Enzym Urease, welches im Boden aber auch in Pflanzen und zahlreichen Mikroorganismen vorkommt (AMBERGER 1996). Im Rahmen eines BMBF-Verbundprojektes wurde geprüft, welchen Einfluss neue, synthetische Ureaseinhibitoren auf die Verwertung des Harnstoff-N haben. Dazu wurden zwei Gefäß- und ein Parzellenfeldversuch(e) mit Hafer bzw. Sommerweizen durchgeführt. Neben Kontrollen, welche nur Harnstoff erhielten, wurden Varianten angelegt, auf die verschiedene Versuchspräparate sowie ein kommerziell erhältliches Präparat gemeinsam mit Harnstoff appliziert wurden.

Folgende Ergebnisse konnten erzielt werden:

1. Im Vergleich zur Kontrolle waren bei Hafer sowohl bei Zwischenernten im vegetativen Stadium als auch bei der Ernte zur Reife deutliche Zuwächse beim Gesamtertrag nach Inhibitoranwendung zu verzeichnen.

2. Stickstoff- und Chlorophyllgehalte (im Stadium des Rispenschiebens bestimmt) wiesen ebenfalls eine Erhöhung auf, die sich entsprechend im erhöhten N-Ertrag niederschlugen.

3. Betrachtet man die Korn- bzw. Kolbenerträge bei Hafer und Mais, so ist eine signifikante Erhöhung bei Anwendung des neuen Versuchspräparates zu Hafer feststellbar, im Falle von Mais ergab das Vergleichspräparat gesicherten Mehrertrag.

4. Zunahmen waren auch bei den Haferstroherträgen zu verzeichnen, ließen sich aber nicht statistisch sichern.

5. Die Auswertung des Parzellenfeldversuches mit Sommerweizen ergab Erhöhungen im Falle des Kornertrages von 8 – 9 % bei deutlichen Steigerungen der Rohproteinerträge.

6. Bei geteilten Stickstoffgaben ist offensichtlich nur die Inhibitoranwendung zur ersten N-Gabe erfolgversprechend.

Eingehende Untersuchungen zur Gesamtbilanzierung des verabreichten N im System Pflanze-Boden-Atmosphäre stehen noch aus.

Literaturverzeichnis

AMBERGER, A., 1996: Pflanzenernährung (Plant Nutrition). 4th Edition. Uni-Taschenbuch 846, Verlag Eugen Ulmer, Stuttgart.

Prozessregulation in der Rhizosphäre
13. Borkheider Seminar zur Ökophysiologie des Wurzelraumes
Hrsg.: W. Merbach, B. W. Hütsch, L. Wittenmayer, J. Augustin
B. G. Teubner Stuttgart, Leipzig, Wiesbaden (2003), S. 138-144

Methodik zu Quantifizierung des Eintrages von Wurzelzellwandresten in den Boden während des Wachstums von Maispflanzen

Jürgen AUGUSTIN[1], Jörg PLUGGE[1], Jürgen PÖRSCHMANN[2], Rainer REMUS[1] und Katja HÜVE[1]
[1]Institut für Primärproduktion und Mikrobielle Ökologie, ZALF e.V., Eberswalder Str. 84, D-15374 Müncheberg,;
[2]Umweltforschungszentrum Leipzig-Halle GmbH (UFZ), Sektion Sanierungsforschung, Permoserstr. 15, D-04318 Leipzig

Abstract

Root structural residues are one of the main sources for organic C input into the soil. However, little is known about the actual input of root cell wall residues and relevant fractions (cellulose, hemicellulose, lignin) into soil, especially during the ontogenesis of plants. In order to solve this problem we combined a repeated pulse labelling of shoots by $^{14}CO_2$ for quantification rhizosphere ^{14}C fluxes with procedures to characterize root residues and selected fractions from root residues (lignin, carbohydrates) at the example of maize. The C flux via root decay during ontogenesis amounted to approx. 116 kg C per ha, approx. 20% of the total C input into the soil as a result of net root growth. In these root residues were 30 kg C as cellulose, 13 kg C as hemicellulose, and 24 kg C as lignin. While the root fractions of old plants showed a uniform ^{14}C distribution, there were strong inhomogeneity with young plants in this regard.

Einleitung

Seit langem gelten die in den Zellwandresten von Wurzeln enthaltenen Stoffgruppen wie Cellulose, Hemicellulose und Lignin als wichtige Quellen für die Humusneubildung (SIMONART und MAYAUDON 1966, HAIDER 1996). Möglicherweise ist ihre reale Bedeutung bislang aber unterschätzt worden, fanden doch bei solchen Betrachtungen zumeist nur die zum Zeitpunkt des Absterbens der Pflanze im Boden vorhandenen Wurzelreste Berücksichtigung (JENKINSON 1990, KÖRSCHENS und SCHULZ 1999). Tatsächlich kommt es schon sehr frühzeitig im Verlauf der Ontogenese zur Entstehung und zur mikrobiellen Umsetzung von Wurzelzellwandresten. So starben z. B. beim Mais die Wurzelhaare nach jeweils 2 bis 3 Tagen ab, während dies für Lateralwurzeln ab dem 6-Blattstadium zutraf (FUSSEDER 1987). Unmittelbar danach beginnt die mikrobielle Verwertung sowohl der zellulosehaltigen Strukturen als auch der mit Polyphenolen imprägnierten, d. h. der ligninhaltigen Mittellamellen der Zellwandreste (FOSTER und ROVIRA 1976). Die Vorstellungen zum Umfang der im Verlauf der Ontogenese anfallenden Wurzelreste bzw. der darin enthaltenen Mengen an hochmolekularen Verbindungen mit besonderer Relevanz für die Bildung stabilisierter organischer Bodensubstanz sind jedoch noch immer sehr vage.

Zielstellung eigener Untersuchungen war es daher, im Kontext mit einem auf der ^{14}C-Tracertechnik basierenden Verfahren zur Bestimmung der Assimilat-C-Flüsse in der Rhizosphäre von Maispflanzen (AUGUSTIN *et al.* 2002) eine Methode zur Bestimmung der in den Wurzelresten enthaltenen Stofffraktionen zu entwickeln. Die Kombination der Resultate beider Verfahren sollte eine Abschätzung der Mengen gestatten, mit denen relevante Stoffgruppen aus den Wurzelzellwandresten im Verlauf der Ontogenense in den Boden eingetragen werden. Zugleich galt es mit Hilfe dieser Prozedur zu prüfen, ob denn die angewandte ^{14}C-Impulsmarkierungstechnik zur einer homogenen ^{14}C-Markierung der einzelnen Wurzelfraktionen geführt hat. Nur wenn dies gegeben ist, können die mit Hilfe der genannten Isotopenmethode erzielten Resultate zu den Assimilat-C-Flüssen als vertrauenswürdig und korrekt angesehen werden (SWINNEN *et. al* 1994, AUGUSTIN *et al.* 2002).

Material und Methoden

Zur Bestimmung der unterirdischen Assimilat-C-Flüsse wurden immer neue Maispflanzen, die unter kontrollierten Bedingungen in Mitscherlichgefäßen aufwuchsen, zu unterschiedlichen, über die gesamte Ontogenese verteilten Zeitpunkten einer ^{14}C-Impulsmarkierung unterzogen. An jede dieser Markierungen schloß sich über einen Zeitraum von ca. drei Wochen die Erfassung von Veränderungen in der ^{14}C-Verteilung im System Pflanze-Boden an. Aus den Resultaten dieser Untersuchungen ließen sich dann mit Hilfe von Modellen zum Pflanzenwachstum die kumulierten Assimilat-C-Flüsse im System Pflanze-Boden während der Ontogenese errechnen. Die Menge der während des Pflanzenwachstums absterbenden Wurzeln (synonym: Wurzelzerfall) wurde als Differenzgröße zwischen dem Brutto-C-Eintrag in die Wurzelstrukturbestandteile und dem tatsächlich gemessenen Wurzelzuwachs bestimmt (AUGUSTIN *et al.* 2002). Weitere Details zur Versuchsdurchführung: Hybridmais der Sorte „Saphir", S210, auf Krumenmaterial vom **Versuchsstandort Kühnfeld/Halle** bei 60% der maximalen Wasserkapazität unter kontrollierten Bedingungen im Mitscherlichgefäß angezogen (Temperatur 21°C, Beleuchtung über 12 h mit Photonenfluß von 300 μmol m^{-2} s^{-1}, relative Luftfeuchte 90%). Die ^{14}C-Impulsmarkierung und die Bestimmung der

^{14}C-Aktivitäten im CO_2 bzw. in den Boden- und Pflanzenfraktionen erfolgte nach den bei RICHERT *et al.* (2000) beschriebenen Prozeduren.
Zur Charakterisierung der in Maiswurzeln (aus dem Boden ausgewaschen und getrocknet) enthaltenen Stoffgruppen wurde – basierend auf Arbeiten von ALLEN *et al.* (1974) - ein Fraktionierungsschema entwickelt und validiert, das eine Auftrennung in Lipide, einen heißwasserlöslichen Anteil, Cellulose, Hemicellulose und Lignin erlaubt (Abb. 1). Hierbei erfolgte als erster Schritt die Befreiung des „polymeren" Wurzelanteils von den reversibel sorbierten, niedermolekularen Substanzen. Im Anschluß daran ließen sich dann die Zellwandpolymere in Cellulose, Hemicellulose sowie Lignin auftrennen (Abb. 1).

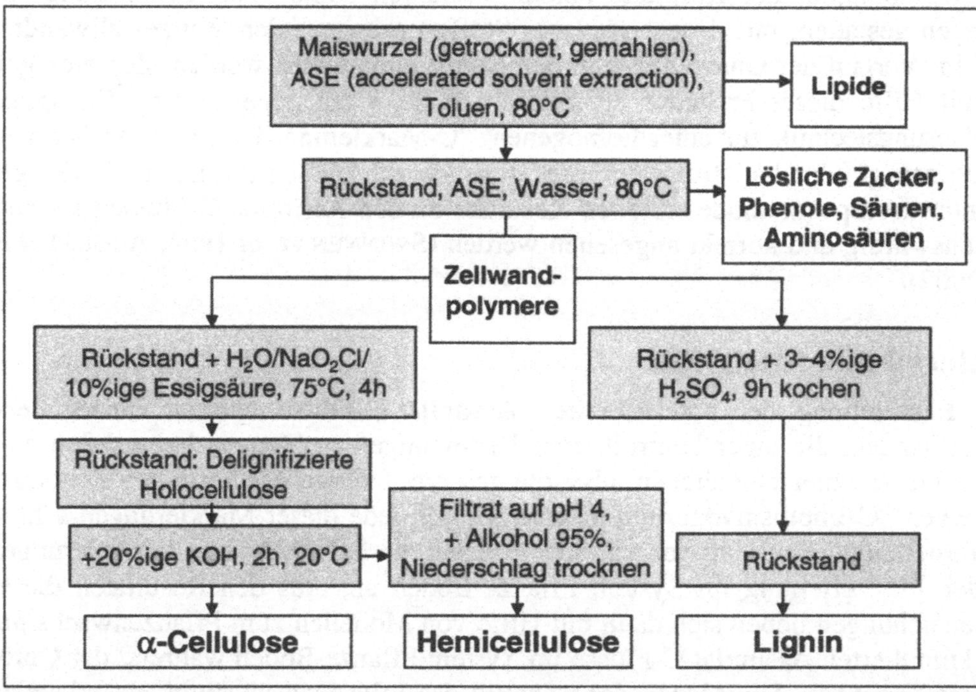

Abb. 1: Vereinfachtes Schema der Wurzelfraktionierung (nach ALLEN *et al.* 1974, erweitert und modifiziert)

Die Analyse der Fettsäuremuster der Lipide erfolgte mittels alkalischer Verseifung und GC/MS-Analyse der Methylester. Der Heißwasserextrakt wurde mittels HPLC in Stoffgruppen vorgetrennt, sowie mittels Elektrospray-MS und NMR-Spektroskopie näher untersucht. Die Bestimmung der Cellulose und Hemicellulose erfolgte auf Grundlage von Ionenchromatographie und amperometrischer Detektion nach Freisetzung der Monomeren. Der Ligninanteil ließ sich durch die bekannte Oxidation mit CuO näher charakterisieren, wobei der nach dem Aufschluss gewonnene organische Extrakt methyliert wurde (ALLEN 1974, GORECKI und PÖRSCHMANN 2001, PÖRSCHMANN *et al.* 2001). Zuletzt wurde mit Hilfe des bei RICHERT *et al.* (2000) beschriebenen Verfahrens in jeder Fraktion die ^{14}C-Aktivität ermittelt.

Ergebnisse und Diskussion

Insgesamt gestattete das Fraktionierungsverfahren eine relativ präzise und verlustarme Auftrennung der trockenen Wurzelreste (Tab. 1). Der Heißwasserextrakt war sehr komplex aus verschiedenen niedermolekularen Verbindungen zusammengesetzt. Im Fettsäuremuster fand sich eine typische even-over-odd Diskriminierung mit charakteristischen Anteilen monoungesättigter Fettsäuren (Daten nicht dargestellt, PÖRSCHMANN *et al.* 2001). Erwartungsgemäß hatten aber die aus Zellwänden stammmemdem Stofffraktionen (Cellulose, Hemicellulose und Lignin) den größten Anteil an der Wurzeltrockenmasse. Im Verlauf der Ontogenese ist es aber offensichtlich zu deutlichen Veränderungen in der Zusammensetzung der Wurzelreste gekommen. So sank der Anteil der heißwasserlöslichen Fraktion und der Hemicellulose im Verlauf der Ontogenese deutlich ab, während der Anteil von Cellulose und Lignin anstieg bzw. auf hohem Niveau verharrte. Mit hoher Wahrscheinlichkeit ist das auf einem zum Ende des Pflanzenwachstums

Tab. 1: Anteil der Stofffraktionen an der Trockenmasse von Maiswurzeln; (Angaben in % der Trockenmasse, die Gesamtausbeute beträgt 82 bis 90%; die Bestimmung erfolgte mit Hilfe des Verfahrens nach ALLEN *et al.* 1974)

Fraktion	**Pflanzenalter**			
	36 d	**50 d**	**91 d**	**Mittelwert**
Lipide	1,2	4,8	2,1	2,7
Heißwasserextrakt	19,0	14,0	9,3	14,1
Cellulose	26,0	26,3	24,8	25,7
Hemicellulose	20,0	10,1	4,7	11,6
Lignin	16,0	24,7	20,4	20,4

Tab. 2: Verteilung der spezifischen ^{14}C-Aktivität in den Wurzelstofffraktionen (Angaben in Bq mg^{-1} Stoff-C)

Fraktion	**Pflanzenalter**		
	36 d	**50 d**	**91 d**
Lipide	-	288	114
Heißwasserextrakt	-	729	210
Cellulose	5010	432	112
Hemicellulose	1117	250	110
Lignin	4210	320	124

hin abnehmenden Anteil lebender, mit Cytoplasma gefüllter Zellen zurückzuführen.
Um die mit Wurzelwachstum und –zerfall in Zusammenhang stehenden C-Flüsse zu veranschaulichen, wurden die im Modellexperiment ermittelten Resultate unter Bezugnahme auf den C-Fluss in den Spross in Angaben pro ha umgerechnet. Dementsprechend lägen, wenn von einer für mittlere Silomaiserträge typischen Spross-C-Menge von 4000 kg pro ha ausgegangen wird, die kumulierten C-Flüsse für das Nettowurzelwachstum bei 508 kg C pro ha und für den Wurzelzerfall während der Ontogenese bei ca. 116 kg C pro ha (AUGUSTIN *et al.* 2002). Mit anderen Worten: Falls der bereits im Verlauf der Ontogenese einsetzende Wurzelzerfall keine Berücksichtigung gefunden hätte, wäre der von Maiswurzeln bewirkte C-Eintrag in den Boden um ca. 20% unterschätzt worden. Angesichts dessen erscheint es sinnvoll, in nachfolgenden Labor- und Freilandexperimenten (auch an anderen Pflanzen) die tatsächliche Rolle der während des Pflanzenwachtums anfallenden Wurzelreste als C-Quelle systematisch zu klären. Das gilt umso mehr, als, ausgehend von o. g. Relationen und den in Tabelle 1 angegebenen Mittelwerten, auf diesem Wege wohl tatsächlich pro ha bereits ca. 30 kg Cellulose-C, 13 kg Hemicellulose-C und 24 kg Lignin-C in den Boden gelangt sind. Ungeachtet der zunächst sehr gering erscheinenden Mengen könnte dies für die Bildung stabilisierter organischer Substanz von erheblicher Relevanz sein, sollen doch wurzelbürtige Makromoleküle sehr viel stärker in die Huminsäure- bzw. Huminfraktion eingebaut werden als solche aus oberirdischen Pflanzenteilen (FOKIN 1978, MARTIN und MERCKX 1992). Ob das auch für Maiswurzeln zutrifft, müsste jedoch ebenfalls in weitergehenden Untersuchungen überprüft werden.
Einen sehr starken Einfluss hatte das Pflanzenalter auch auf die spezifischen ^{14}C-Aktivitäten in den Wurzelfraktionen. Je älter die markierten Pflanzen waren, um so deutlicher ging die spezifische ^{14}C-Aktivität zurück (Tab. 2). Ursache dafür war wahrscheinlich die zunehmende Verdünnung des mit gleichbleibender Gesamtaktivität applizierten ^{14}C-Tracers durch die wachsende Menge an unmarkierten Pflanzen- bzw. Wurzel-C. Sehr differenziert fielen hingegen die Resultate zur Homogenität der ^{14}C-Markierung der Wurzelfraktionen aus. Während bezüglich der spezifischen ^{14}C-Aktivität zwischen den Fraktionen anfänglich große Unterschiede bestanden, kam es zum Ende des Wachstums hin zu einer nahezu vollständigen Angleichung (Tab. 2). Die inhomogene ^{14}C-Verteilung zu Vegetationsbeginn ist möglicherweise auf die zu dieser Zeit stark ausprägten Unterschiede in der Dynamik des C-Stoffwechsels der einzelnen Fraktionen zurückzuführen. Aus diesen Befunden ergibt sich zunächst, dass im Gegensatz zur Auffassung von SWINNEN *et al.* (1994) **zumindest bei jungen Pflanzen mehr als drei Wochen notwendig sind, um eine vollkommen homogene Verteilung der ^{14}C-Aktivität in den einzelnen Wurzelfraktionen zu erreichen.** Andererseits sollten sich aus

der Möglichkeit, die Stofffraktionen in den Wurzeln und ihre spezifische ^{14}C-Aktivität direkt bestimmen zu können, auch Ansatzpunkte ergeben (z. B. Korrekturrechnungen), die ungeachtet inhomogener ^{14}C-Verteilungen eine präzise Erfassung der Assimilat-C-Flüsse mit Hilfe der ^{14}C-Impulsmarkierung im Boden gestatten. Dies ausführlich zu testen, bleibt gleichfalls späteren Experimenten vorbehalten.
Zusammenfassend läßt sich feststellen, dass die Kombination von ^{14}C-Tracertechnik und Wurzelfraktionierung wahrscheinlich nicht nur bei der Bestimmung des Umfanges der während der Ontogenese anfallenden Wurzelreste und der darin enthaltenen hochmolekularen Verbindungen, sondern auch bei der exakten Quantifizierung aller Assimilat-C-Flüsse im Boden neue Perspektiven eröffnet. Endgültige Aussagen hierzu sind aber erst nach wesentlich umfassenderen Untersuchungen möglich.

Danksagung

Die vorgestellten Untersuchungen wurden in Rahmen des Schwerpunktprogramms 1090 „Böden als Quelle und Senke für CO_2" von der Deutschen Forschungsgemeinschaft gefördert. Spezieller Dank gilt Frau Mirus, Frau Herrendorf, Herrn Blasinski (ZALF e. V.) sowie Frau Becker (UFZ GmbH) für die vorzügliche technische Unterstützung.

Literatur

ALLEN, S. E. *et al.* (1974): Chemical analysis of ecological materials. Blackwell scientific publications, Oxford.

AUGUSTIN, J., PLUGGE, J., PÖRSCHMANN, REMUS, R., HÜVE, K., HÜTSCH, B. W., MERBACH , W., (2002): Methodik zur Quantifizierung des Eintrages von Wurzelzellwandresten in den Boden während des Wachstums von Maispflanzen. In: Merbach, W.; Hütsch, B.; Wittenmayer, L.; Augustin, J. (Hrsg.) *Durchwurzelung, Rhizodeposition und Pflanzenverfügbarkeit von Nährstoffen und Schwermetallen.* 116-123, Stuttgart, Leipzig (Teubner).

FOKIN, O. A. (1978): Sravnitelnoe izuchenie uchastija nazemnyh i konevykh rastilnykh ostakov v formirovani i obnovlenü pochvennogo gumusa. *Tezisy dokladov 5. delegatskogo sezda vses. olscesha poc*, Minsk, 45-46.

FOSTER, R. C., ROVIRA, A. D. (1976). The ultrastructure of wheat rhizosphere. *New Phytologist* 76, 343-352.

FUSSEDER, A. (1987): The longevity and activity of the primary root of maize. *Plant and Soil* 101, 257-265.

GÓRECKI, T., PÖRSCHMANN, J. (2001): In-Column Pyrolysis – A New Approach to an Old Problem. *Anal. Chem.*, 73, 2012-2017.

JENKINSON, D. S. (1990): The turnover of organic carbon and nitrogen in soil. *Philosophical Transactions of the Royal Society., B.* 329, 361-368.

KÖRSCHENS, M., SCHULZ, E. (1999): Die organische Bodensubstanz, Dynamik - Reproduktion - ökonomisch und ökologisch begründete Richtwerte. UFZ-Bericht 13/1999, Umweltforschungszentrum Leipzig-Halle GmbH.

MARTIN, J. K., MERCKX, R. (1992): The partitioning of photosythetically fixed carbon within the rhizosphere of mature wheat. *Soil Biol. Biochem.* 24, 1147-1156.

PÖRSCHMANN, J., PLUGGE, J., TOTH, R. (2001): In-situ Derivatisation using Pressurized Liquid Extraction to Determine Phenols, Sterols, and Carboxylic Acids in Environmental Samples and Microbial Biomasses, *J. Chromatogr. A,* 909, 95-109.

RICHERT, M., SAARNIO, S., JUUTINEN, S., SILVOLA, J., AUGUSTIN, J., MERBACH, W. (2000): Distribution of assimilated carbon in the system *Phragmites* australis-waterlogged peat soil after carbon-14 pulse labelling. *Biol. Fertil. Soils*, 32, 1 – 7.

SIMONART, P., MAYAUDON, J. (1966): Etude des transformations de la matière organique du sol au moyen du carbone-14. In: The use of isotopes in soil organic matter studies. Special supplement to the Journal of applied radiation and isotopes, Pergamon Press, Oxford, 245-258.

SWINNEN, J., VAN VEEN, J. A., MERCKX, R. (1994): Rhizosphere carbon fluxes in field grown spring wheat: model calculations based on ^{14}C partitioning after pulse-labelling. *Soil Biol. Biochem.* 26, 171-182.

Verzeichnis der Teilnehmer

Prof. Dr. Jürgen Augustin, Institut für Primärproduktion und Mikrobielle Ökologie im ZALF Müncheberg, Eberswalder Str. 84, D-15374 Müncheberg

Astrid Appuhn, Gesamthochschule Kassel, Fachgebiet Bodenbiologie, Nordbahnhofstr. 1a, D-37213 Witzenhausen

Henry Barus, Universität Hohenheim, Institut für Pflanzenernährung, Fruwirthstr. 20, D-70599 Stuttgart

Dr. Heidrun Beschow, Institut für Bodenkunde und Pflanzenernährung der Martin-Luther-Universität Halle-Wittenberg, Adam-Kuckhoff-Str. 17b, D-06108 Halle/Saale

Anne Bolhöfer, Agrikulturchemisches Institut der Rheinischen Friedrich-Wilhelms-Universität Bonn, Karlrobert-Kreiten-Str. 13, D-53115 Bonn

Annelies Claus, Institut der Pflanzenbauwissenschaften, Humboldt-Universität Berlin, Invalidenstraße 42, D-10115 Berlin

Dr. Annette Deubel, Institut für Bodenkunde und Pflanzenernährung der Martin-Luther-Universität Halle-Wittenberg, Adam-Kuckhoff-Str. 17b, D-06108 Halle/Saale

Frauke Diekmann, GSF-Forschungszentrum für Umwelt und Gesundheit, Ingolstädter Landstr. 1, D-85764 Neuherberg

Kay Domeyer, Agrikulturchemisches Institut der Rheinischen Friedrich-Wilhelms-Universität Bonn, Karlrobert-Kreiten-Str. 13, D-53115 Bonn

Dr. Wolfgang Gans, Institut für Bodenkunde und Pflanzenernährung der Martin-Luther-Universität Halle-Wittenberg, Adam-Kuckhoff-Str. 17b, D-06108 Halle/Saale

Stephan Gantner, GSF-Forschungszentrum für Umwelt und Gesundheit, Ingolstädter Landstr. 1, D-85764 Neuherberg

Prof. Dr. Eckard George, Institut für Gemüse- und Zierpflanzenbau, Großbeeren/Erfurt e.V. (IGZ), Theodor-Echtermeyer-Weg 1, D-14979 Großbeeren

Dr. Sylvie Herrmann, Universität Jena, Institut für Ökologie, Lehrbereich Umweltwissenschaften, Dornburger Str. 159, D-07743 Jena

Volker Hoffmann, Institut für Bodenkunde und Pflanzenernährung der Martin-Luther-Universität Halle-Wittenberg, Adam-Kuckhoff-Str. 17b, D-06108 Halle/Saale

Dr. Michael Kaldorf, Universität Jena, Institut für Ökologie, Lehrbereich Umweltwissenschaften, Dornburger Str. 159, D-07743 Jena

Dr. Mahmoud Kamh, Universität Hannover, Institut für Pflanzenernährung, Herrenhäuser Straße 2, D-30419 Hannover

Dr. Bärbel Kiesel, UFZ, Sektion Umweltmikrobiologie, Permoserstr. 15, D-04318 Leipzig

Annette Krechel, Institut für Molekulare Physiologie und Biotechnologie, Mikrobiologie, Universität Rostock, Albert-Einstein-Str. 3A, D-18051 Rostock

Dr. Rolf Kuchenbuch, Institut für Primärproduktion und Mikrobielle Ökologie im ZALF Müncheberg, Eberswalder Str. 84, D-15374 Müncheberg

Dr. Yakov Kuzyakov, Universität Hohenheim, Institut für Bodenkunde und Standortslehre (310), D-70593 Stuttgart

Prof. Dr. E. Leistner, Institut für Pharmazeutische Biologie der Rheinischen Friedrich-Wilhelms-Universität Bonn, Nussallee 6, D-53115 Bonn

Andreas Lieber, UFZ, Sektion Umweltmikrobiologie, Permoserstr. 15, D-04318 Leipzig

Prof. Dr. Wolfgang Merbach, Institut für Bodenkunde und Pflanzenernährung der Martin-Luther-Universität Halle-Wittenberg, Adam-Kuckhoff-Str. 17b, D-06108 Halle/Saale

Ilka Meinen, Agrikulturchemisches Institut der Rheinischen Friedrich-Wilhelms-Universität Bonn, Karlrobert-Kreiten-Str. 13, D-53115 Bonn

Henrike Perner, Institut für Gemüse- und Zierpflanzenbau, Großbeeren/Erfurt e.V. (IGZ), Theodor-Echtermeyer-Weg 1, D-14979 Großbeeren

Dr. Martina Puhlmann, UFZ, Sektion Bodenforschung, Theodor-Lieser-Str. 4, D-06120 Halle

Anne-Christine Schmidt, UFZ Leipzig-Halle GmbH, Sektion Analytik, Theodor-Lieser-Str. 4, D-06120 Halle

Svea Schülter, Agrikulturchemisches Institut der Rheinischen Friedrich-Wilhelms-Universität Bonn, Karlrobert-Kreiten-Str. 13, D-53115 Bonn

Mika Tarkka, Botanisches Institut, Physiologische Ökologie der Pflanzen, Eberhard-Karls-Universität Tübingen, Auf der Morgenstelle 1, D-72076 Tübingen

Adel Rabie Usman, Universität Hohenheim, Institut für Bodenkunde und Standortslehre (310), D-70593 Stuttgart

Susanne Wings, Institut für Pharmazeutische Biologie der Rheinischen Friedrich-Wilhelms-Universität Bonn, Nussallee 6, D-53115 Bonn

Dr. Lutz Wittenmayer, Institut für Bodenkunde und Pflanzenernährung der Martin-Luther-Universität Halle-Wittenberg, Adam-Kuckhoff-Str. 17b, D-06108 Halle/Saale

Autorenregister

Sachregister

GPSR Compliance

The European Union's (EU) General Product Safety Regulation (GPSR) is a set of rules that requires consumer products to be safe and our obligations to ensure this.

If you have any concerns about our products, you can contact us on ProductSafety@springernature.com

In case Publisher is established outside the EU, the EU authorized representative is:

Springer Nature Customer Service Center GmbH
Europaplatz 3
69115 Heidelberg, Germany

Batch number: 10613743

Printed by Printforce, the Netherlands